THE DARK SIDE OF TECHNOLOGY

THE DARK SIDE OF TECHNOLOGY

Peter Townsend

OXFORD
UNIVERSITY PRESS

Great Clarendon Street, Oxford, OX2 6DP,
United Kingdom

Oxford University Press is a department of the University of Oxford.
It furthers the University's objective of excellence in research, scholarship,
and education by publishing worldwide. Oxford is a registered trade mark of
Oxford University Press in the UK and in certain other countries

First Edition published in 2016

Impression: 1

Published in the United States of America by Oxford University Press
198 Madison Avenue, New York, NY 10016, United States of America

British Library Cataloguing in Publication Data
Data available

Library of Congress Control Number: 2016939460

ISBN 978–0–19–879053–2

Printed and bound by
CPI Group (UK) Ltd, Croydon, CR0 4YY

Acknowledgement

It is a great pleasure to thank my friend Angela Goodall for her continuous, supportive, critical, and imaginative input during the many drafts involved in preparing this book.

Contents

1

Have We the Knowledge, Willpower, and Determination to Survive?

Humans are an intelligent and aggressive species that uniquely uses language and writing to pass experience and skills from one generation to the next. This enabled technological advances to develop as we moved from caves to the modern world. Technology is the keystone to our progress, from flint tools to metals, transport, and electronics. It underpins our immense progress in agriculture, biology, and medicine. The positive results are hugely beneficial. For many, we have food, luxuries, and communication on a global scale, and live longer.

All this is wonderful and true progress, but there is a dark side to technology, as the very same knowledge has brought better weapons, warfare, and oppression. We destroy forests, our resources, and other creatures. We ignore their extinctions in our rush forward for personal or corporate gain. Our actions are frequently blinkered and self-centred, even in farming and fishing, which we need for our survival.

If we are to overcome the problems we are causing, we must first identify and recognize them. Progress has relied on transmitting information, but the ways we record and store it are changing so rapidly that past knowledge is lost. Stone carvings last thousands of years, but Internet communications are lost or deleted in minutes. The formats of our hardware and software evolve so quickly they become obsolete within a decade; photographs have survived for two centuries, but electronic images are ephemeral. In agriculture, we put our faith in specialized monoculture crops, and ignore the possibility of new diseases that can eradicate them.

Computer and communication advances are unquestionably beneficial, but they have opened up opportunities for cybercrime, malicious access to electronically stored data, lack of privacy, and a powerful weapon for warfare and terrorism. Damage is simple to implement. There are also downsides for people who cannot cope with such new technologies.

Globalization and communications are marvellous for reaching other cultures and disseminating information or new styles of music, but equally they are destroying ancient languages at an unprecedented rate. Their loss also destroys the culture and knowledge that was within them. Even rapid global transport has a dark side, as not only people but also new pandemics can cross the world in a day.

Reliance on advanced, interconnected, sophisticated technology has made us vulnerable to natural and manufactured events that previously were unimportant. For example, loss of communication systems and electrical supplies would bring prolonged chaos and anarchy, as 'civilized' life is now absolutely dependent on them. All the major developed countries are at risk from such natural phenomena.

The book aims to promote a realization of how our lives have been altered and controlled, even in subtle ways, by the advances in science and medicine. Progress has been excellent, and will continue to be so, as long as we capitalize on the positive aspects and avoid the others. For example, food supplies are varied and increasing, but lack of self-control brings obesity and many related diseases, as a direct result of easy availability.

Scientific advances have brought longevity, health, and affluence for many of the advanced nations, plus an exponential growth in population, especially in the underdeveloped ones. We are at a threshold where demands for resources and food are already beyond a sustainable level if we were to raise living conditions across the world to that of the developed, richer nations. This spells trouble, because underdeveloped nations want to better their conditions, and this is a justifiable objective.

Overpopulation and food shortages are insidious, but there are many other disaster scenarios that could occur rapidly and unpredictably. They may even happen in the very near future and, just as for the improvements we are enjoying, they are the result of our reliance on and obsession with technological advance.

We like disaster movies, so in Chapter 2 I will outline a film script to show that natural and frequently occurring events, irrelevant for past generations, have the potential to wipe out technologically advanced societies. We have the knowledge to anticipate such events and can make contingency plans to survive. This is a positive and essential approach that needs to be actively started immediately. I will therefore suggest what actions are needed. Failure to be prepared will not be the end of humanity, but it could easily destroy many advanced nations. World power would then shift to the currently Third World regions.

We can overcome many of these future difficulties if we have sufficient information, data, knowledge, and understanding, but knowledge and data now vanish ever more quickly in new storage formats. Similarly, we need to appreciate that our progress is driven by the young, but their focus on their generation can mean that older people are effectively more isolated than they would be in a non-technological societies.

My hope is that by exposing and discussing the problems generated by technology, we can recognize them and attempt to solve them. Indeed we must.

2

Technology and Survival—Are They Compatible?

Disaster movie scenarios

Civilization depends on science and technology, in every moment and aspect of our lives from energy to food, transport, communication, and knowledge. We need technology to survive, but are increasingly so reliant on it that without it, our civilization would collapse. There are many weak points where we are incredibly vulnerable, or where technologies are actually generating seeds of disaster. I will highlight a number of these, but to set the flavour of the problems that exist, I will first consider the chaos and deaths that would happen if large sections of the world suddenly lost electrical power for a sustained period, and therefore all the systems dependent on it.

Our first reaction is probably to only worry that this means a sudden loss of facilities based on our advanced electronics and computing power. Certainly this would be serious, but in fact it would only be a small part of a global catastrophe. Clearly, I have the essence of a plot for a disaster movie. The only difference is that in the movie version, the world would be saved at the last moment by some brilliant, handsome or beautiful, multi-skilled scientist.

Reality is different. Some scientists may well be handsome or beautiful, but finding a multi-skilled scientist at exactly the right location would be difficult, and in my scenario there will be no easy escape. I am going to pick a very realistic, feasible, and highly probable cause for my movie plots. My concern, and the very reason I have picked this lurking danger, is not that it is a highly unlikely, but, on the contrary, it is one that is certain to happen. Only the magnitude is unpredictable, so I can have two movie plots: one for a modest and one for a severe event. Most successful disaster movies have a sequel, but in my scenarios either or both could occur.

If I consider only a realistic natural-disaster script, I can ignore the really big events, such as large meteorite impacts, which killed off the dinosaurs and most other species. They are fantastically devastating; they have happened, and may well occur again, but our chances of survival are so remote that there is no way we can plan how to cope. After the impact, we will be powerless, and dead. The good news is that they are extremely rare.

The far more interesting considerations, and the ones we should prepare for, are of major catastrophes, or dangers from existing conditions and events, which are certain to happen within the foreseeable future. Humans are quite arrogant and self-centred, so we tend to assume that because we have a highly developed society, and have survived for a few tens of thousands of years, we will always overcome natural disasters. Optimism is great, but my pragmatic view is very different as I realize that, precisely because of our dependence on advanced technologies, our civilization is far more likely to be destroyed by catastrophes that would have been only minor features for earlier generations. These are not science fiction–type scenarios (e.g. aliens or monsters) as there are a sufficient number of probable phenomena where we have built ourselves a technological Achilles' heel.

Who will be vulnerable?

I have picked a natural event which, although common, was totally unimportant for our survival, and in no way was a hazard in the age before our total reliance on electronics. The first hints of danger occurred in the 1850s and 1920s when electrical storms damaged the electrical systems of their time. Neither were major features, as they destroyed just isolated and localized items of simple electrical equipment. Indeed, I suspect we would have easily survived such a natural event just 30 years ago before our modern computer, mobile phone, and Internet society. Therefore all of us are vulnerable if we live in societies where these are key elements of daily life. Unfortunately, it means I am not offering a mythical film plot, but outlining an incredibly serious and potential catastrophe. The positive feature is that it is still in the future and we are able to recognize our vulnerability, so we can actively plan and make suitable defensive preparations to minimize the damage.

Therefore those of us living in high-technology regions may be at risk, whereas those in underdeveloped areas will not immediately feel the consequences of my electrical storm damage.

The villain

The villain behind our danger is the sun. It gives us heat, light, and energy, and therefore life. The solar power source is a massive nuclear reactor with plasma of incredibly high temperature within it, and a surface of very hot ionized gases swirling around, and sometimes erupting from the surface. With a telescope, and eye protection to block the light from the main solar disc, it is easy to see flares leaping from the surface of the sun, particularly from the darker sunspots. Not only is there light from the flares, but also, large amounts of hot ionized gas are continually being ejected from the surface. Our solar science is good, so we can understand why this happens, and how magnetic fields bend the plasma material into loops and cause the explosions off the sun's surface. We even know that there is periodicity in terms of how many sunspots occur: they peak roughly every 11 years. We also know that the spots spread across the surface in a recognizable fashion. We are happy that they occur, because the heat, visible light, UV light, and X-rays that are emitted seem essential for running the chemistry of Earth's atmosphere and our lives. That was the good news—now for the disaster scenarios.

The surface activity on the sun emits flares of ionized material and large ejections of material (called coronal mass ejections). They are a little like massive flamethrowers as we can see the light (plus heat and X-rays) emitted in all directions, and the fuel of the flamethrower is mostly moving in a forward direction. My description is too bland and not scary enough for a disaster movie. I have not yet mentioned the scale of the events. These sunspot flamethrowers become truly terrifying once we know that a typical sunspot has a diameter comparable to the size of the earth. One of them in 2014 was ten times our size. So in terms of flamethrowers, this is quite impressive.

These coronal emission events are happening all the time; fortunately, they are fairly directional, so most of them miss us. We see the light and X-rays from the flares about eight minutes after they happen (they are travelling at the speed of light), but the particles move more slowly and arrive into the range of our solar orbit maybe 18 or more hours later. This is 'slow' compared with the speed of light, but top-speed particles can be moving at 2 million miles per hour. If they are heading towards us, there is little we can do to avoid their effects, but at least we may have some slight advance warning to try to protect our electronics, etc. Bigger flares give less visible light, but a far more intense burst of X-rays. 'Modest' flares have a light plus X-ray energy of, say, one sixth the energy output per second of the sun.

I could add numbers to say that the normal surface temperature of the sun is only about 6,000°C (~10,800°F), and internally the sun is a fusion reactor running at a million or so degrees. The magnetic fields at the surface that drive the emissions are immense by our scales, and the 'modest' flares contain more energy than many million atomic bombs. Unfortunately, numbers on this scale are too far from our experience to convey anything except that fantastic quantities of power are involved.

Sunspots appear to have a broad time pattern of major mass ejections, with small ones happening quite often. Significantly large ones send out big bursts of energy that hit us on average about once every 100 to 150 years, and really supersize ones less often, at a few-hundred-year intervals. In the past all this was unimportant, as the only obvious features were some really colourful effects in the sky. They were exciting, entertaining, and, for the superstitious, significant omens of something. We can forget superstition, but they are certainly omens of potential danger. One may see some humour in the fact that very primitive tribes, and even the Egyptians, invariably have worshipped the sun as the provider of life and food. So they would be correct in saying the sunspot disasters are caused by their god. However, they are precisely the people who would never be influenced directly by the event, and who would have considered it impossible that the sun could have dark spot regions.

Instead of religious omens, we now need to worry about our electronics, as the X-rays and charged particles (the ions) can destroy them. Sensitive electronics are built into all the satellites that are sitting outside our protective atmosphere, and even electronics and electrical power systems that are running on the surface of the earth can be at risk of damage or destruction.

Once particles and ions reach us, the earth's magnetic field bends their paths; as they crash into the earth's atmosphere, they excite the gases in the air. The bending effect is greatest near the magnetic poles, and visually the effect is spectacular at high latitudes. There are waves and sheets of light rippling across the sky as the ions excite colours from the nitrogen and oxygen in our atmosphere. This magnificent light display is the aurora borealis (or Northern Lights) in the northern hemisphere, and the aurora australis (Southern Lights) above Antarctica. Bigger solar ejections produce larger events that can be seen farther from the poles. Linked with the light show are both X-rays directly from the sun and X-rays formed during collisions of the ions. Even for modest aurora events, these changes in X-ray exposure are obvious, so during any plane flight in the polar regions, especially at the time of a flare, we have a measurably higher X-ray radiation exposure than in normal flights.

We have always had this free spectacle from the solar flares, but there are technological features as well. The electrical currents and magnetic fields from the ions and energetic electrons produce electrical signals that can swamp any electrical communication or signals being processed by our electronics. Interference happens across the whole range of electromagnetic wavelengths (including microwave, radio, and visible light). So they produce noise and interference at all the frequencies used for mobile phones, radio and TV, and radar. For electronic equipment directly in line with the stream of particles, not only can the electronic functions be drowned by the electrical noise, but the particles are sufficiently energetic that they can physically destroy the electronic components, either by physical impacts of high-energy ions, or by causing large voltage pulses. Ions may not be heavy, but they can be travelling at speeds up to a few hundred thousand miles per hour, so they have a lot of kinetic energy.

The physical damage is much like a very tiny bullet passing through material. So satellites (or astronauts) parked out in space, unshielded by the earth's atmosphere, are highly vulnerable. I have seen photos of astronauts' helmets where damage tracks have been detected entering one side and emerging on the other! Astronauts have often reported seeing flashes of light from the tracks of particles passing through their eyes. In addition to the very obvious destructive impacts, there will be electronic damage, because many structures act as antennae (as for a radio), and bigger antennae generate bigger signals. The effect is to put a high voltage across our sensitive electronics. For example, if we connected the mains voltage across a mobile phone, there would be a big bang and a dead phone. The solar flare damage is effectively able to do the same thing! Turning off the electronics during the bombardment may help, but there is no guarantee that the systems can be reactivated afterwards.

Satellite loss and air traffic

I will start with a modest disaster movie plot that is highly probable (or, according to many people, absolutely certain to happen). Our most exposed electronics, which are not shielded in any way by the earth's atmosphere and our magnetic field, function in the many hundreds of satellites that are orbiting us in space. They are absolutely essential to transmit signals around the earth for long-range communication, for GPS (global positioning systems), many TV channels, banking credit card transactions, and a part of the Internet and mobile phone networks. They are also the backbone of radar surveillance and military early warning systems. They are crucial in long-range weather forecasting. In Europe the satellite business feeds around 80 million households with information and entertainment, and even around 66 million sites rely on satellites to run the cable connections. They may be costly to build and launch, but the rewards are huge, and estimates of just the satellite-driven entertainment businesses are quoted as being some 300 billion dollars/pounds/euros per year.

Safeguarding satellites is therefore a strategic concern. It is essential that we have enough early warning of sunspot eruptions to try to protect the satellites from solar flare radiation, as well as the ions that arrive later on. Unfortunately, since the X-rays travel at the speed of light, attempting a shutdown in time, after we realize the radiation level is unusually high, is not guaranteed. Nor, if we do turn off most functions, is there certainty that we can restart operation after the sunspot activity has finished.

Nevertheless, we can gain some early warning about the particles that arrive some time after the X-ray burst. Since this may be more damaging to the electronics, advance notice is valuable. The danger has been appreciated, and in order to gain some early warning, an unused, 'ancient' satellite has been moved to act as a sentry to look for solar particle emissions. It is being joined with a second, newly launched probe (the 340-m Deep Space Climate Observatory). Both satellites are much closer to the sun than our communications satellites. So they should offer advance early warning of a big particle flux.

Their location is interesting as they are sitting out at a place called the Lagrange point 1. This region is around 1.5 million kilometres (~1 million miles) from Earth, and it is an ideal site because it minimizes fuel: the reason is that the satellites are balanced by the gravitational pull of the sun in one direction and the earth in the other. The new satellite will detect particles, and so can radio a warning that within 15 to 60 minutes particles will strike us. The satellites are of course potentially vulnerable to damage, so could be a modern equivalent of the early Chinese outposts in Mongolia, where a small number of soldiers in tiny forts used flags to signal that invaders were heading towards China—a noble task, as it would be their last action.

What will happen in the modest sunspot scenario?

Among the people most directly at immediate risk from satellite loss—or malfunction of positional coordinates once the system restarts—will be those in aircraft that are using it on automatic pilot control. The danger can be caused by quite small solar flares, which are not exceptionally rare. As an example, in April 2014 a flare blocked communications and GPS over part of the Pacific. In that case, the air traffic density was very low and so there were no accidents. Had the same event happened over Europe or the eastern USA, the consequences would have been dire.

Satellite shutdown and communication loss between aircraft and ground control is now critical as the trend from improved technology is that both navigation and landings are made automatically, with the position extremely well controlled by GPS. Because of this reliance on electronics, there have been reductions in the number of air traffic control staff. Killing the satellite positioning generates a serious landing problem. Furthermore, many planes will go off course, not least because navigational piloting skills by non-electrical means are no longer seen as essential as they once were, and the training and manual experience will inevitably have become reduced for many types of aircraft. Note, however, that military personnel may be better prepared in this respect.

Rather than make extravagant claims, we can actually estimate the possible scale of the difficulties. Each day from major airports there are as many as 3,000 flights crossing the North Atlantic; additionally, there are around 30,000 European flights. North America has both scheduled and non-commercial flights, with current totals around 10,000 per day. For the main airport hubs, loss of GPS landing would be a very real difficulty, especially for an airport such as Heathrow, which is stacking incoming flights and landing a plane roughly once a minute at peak times.

So a loss of satellite navigation is the initial major hazard. For a true disaster movie version, we only need to add a thick cloud level, darkness, or both, over, say, northern Europe (certainly not that rare); because of the electronic noise from the event there will be a loss of communication between ground and crew. It is not a large step to realize that the onboard plane radio links may have failed, or be blocked as part of the overall solar flare electrical noise. In this level of disaster plot, the passenger death toll will soar, together with many more casualties in the areas where the planes hit the ground. For airports such as those in London, Frankfurt, and New York, the crashes could well involve densely populated urban areas.

Unfortunately, I am definitely not making excessive claims here: we do not need solar intervention to cause problems with air traffic movements because they are also vulnerable to equipment failures (or terrorist attacks) that damage the normal functioning of our electronic communications. To underline our sensitivity to disruption, a very minor computer problem in December 2014 at the main national air traffic control in southern England caused cancellations and repercussion effects of delays for several days. This is despite there being duplicate facilities and many fail-safe features, and no loss of power at the facility. So, relative to satellite communication failure, this was a trivial problem. In safety terms, it was handled well.

Failure of satellites or control centres not only creates risks for airline passengers, but has indirect financial effects as the satellite communication business is worth around a trillion dollars per year. So satellite losses would impact the entire global economy for many months, or years, until new satellites could be built and launched.

A ground-based view of this modest event

Loss of satellites will equally impact ground activities. I have already listed some of the satellite functions; for example, bank transactions with credit cards, as well as many types of communication and satnav systems will fail. Therefore, at a very mundane level, not only telephone, but also financial and other transactions would be blocked. So we would have no ability to make purchases in shops or online, and for many transport systems it would not be possible to buy tickets to gain entry to bridges, toll roads, and the like. A further aspect of improved technology is that we now rely on satnav systems—they have replaced maps. Many drivers will not have paper maps as a back-up if their reception fails. Therefore, if satellite navigation failed, many people would be literally lost. Indeed, since the introduction of satnav technology, fewer people learn how to read a map. This is just one example of a skill that is being lost because of reliance on modern technology.

The less predictable consequence is that because Internet, telephones, and so on are linked into the satellite systems, then loss of one segment will put such an overload strain on the remainder that it is highly likely that there will be an electronic traffic jam—a total blockage of all our electronic communications.

If the satellites survive the electrical storm and can be subsequently reactivated, the whole event is likely to last just a few days, although there would be immense costs in terms of industry and communications. Loss of life would depend primarily on aircraft-related disasters. Overall, this modest solar event is serious, but survivable for the bulk of the countries.

History of auroras and sunspot activity

Having started with modest and frequent examples of solar X-rays and coronal emissions, I will rethink the outcomes that would impact modern society from a more major event. On the positive side, they will give far brighter and spectacular aurora illuminations: in 1859 there was a major solar storm with the aurora seen as far south as Cuba (it is called the Carrington event, because he reported it). In 1859 we had the technology to have long-distance Morse code electrical communication along wires. Long lengths of electrical cable make excellent antenna systems for radio receivers, and were (and are) standard equipment when broadcasting at long wavelengths. The 1859 Morse code telegraph cables were responsive to the solar flare currents formed in the atmosphere. They sucked energy from the atmospheric electrical-magnetic storm and sent pulses of high voltage along the cables. There were reports of sparks and shocks to the telegraph operators and surges in power that were greater than provided by the power units. The electrical storm was serious, even for such primitive electrical systems, as telegraph lines and equipment were made non-functional or destroyed.

By 1921 electrical generators and equipment had become major features of industrial nations. Power units were often linked directly to the equipment they were running, rather than being interlinked into any national power grid. This was fortunate, because there was a modest sunspot event that intersected our orbit and caused problems in the northern hemisphere. As in 1859, the telegraph operators reported high voltages, damaged equipment, and flames coming from the cabling; in both the USA and Sweden, telegraph control buildings were burnt down by the electrically generated fires. In New York, the Central Railroad signalling equipment was destroyed and a fire took out at least one building. Overall, the events were isolated because the power and signal lines that acted as antennae to bring excess power into the systems were relatively short and independent.

Vulnerability of modern interconnected power grid networks

We no longer rely on our own private electrical generators. The major advanced regions of the world, such as North America and Europe, have a highly complex pattern of energy sources in power stations operating with a diverse range of technologies (from solar panels to hydroelectric, wind, coal, gas, and nuclear). Maps of electrical grid distributions over these regions look like complex spiders' webs. Power flow is interconnected across wide regions, and adjusted to optimize performance with changing demands and available supplies. The system is sensitive: even minor change in output from one area can need careful adjustment to maintain full power across the whole region. For predictable daily demand, this is obvious, but less expected is that even the partial solar eclipse across Europe of March 2015 needed to be considered in forward planning, as it reduced solar power generation, which provides around 10 per cent of the electrical energy in central Europe. Economically, power generation is more profitable if there is a minimal spare capacity, so just a 5 or 10 per cent drop in supply at a high speed (i.e. from the moving shadow of the eclipse) cannot be rapidly compensated for by increasing output from other sources.

Mostly the power network works quite well, but nothing is infallible, and there have been numerous widespread power losses as a result of accidents, human errors, and excessive demands. In the last decade or so, greater interconnectivity has resulted in more people being affected by a power loss event. Mostly they have been relatively short term—on the scale of hours, or less than a day—but nevertheless they have caused major disruptions to large numbers of people.

A lightning strike on a substation in Brazil in 1999 caused a chain reaction of grid network collapse that hit 70 per cent of the country and affected 97 million people. Brazil has suffered other widespread electrical failures in 2009, 2011, and 2013. The north-eastern USA had power loss for four days in 2003 as a result of a combination of faults and human error that blacked out some 50 million people. Failures in other regions include one in Java in 2005 that hit 100 million people. So far the greatest number of people involved in a single power loss was from a power shortage in India in 2001, where there was power loss for 226 million. My sample of major incidents from a variety of causes underlines the claim that with ever-increasing demands, these power failure problems will be repeated many times in the future.

Consequences of power grid failures

Immediate consequences of these grid failure events are distinctly unpleasant. People have been trapped in elevators, or been unable to leave high-rise buildings if they are unable to walk down many flights of steps. Typically, underground and other railway systems have trapped people; loss of power has blacked out traffic control and communication systems. Water and fuel pumping systems fail without mains power, and sewage pumping and processing have equally stopped. Mobile phone systems have sometimes survived, but because modern phones need recharging relatively often, a power loss of just two days could be fairly disastrous. In most countries, criminal activities have exploited the possibilities of alarm failure and overloaded police response. Financially, the loss of industry, failure of supplies in shops, etc. have been estimated for the short-term examples I am citing, and the monetary numbers normally are quoted in billions. As a guide to the financial impact, the USA has an average of nine hours of power loss per consumer each year as a result of very minor local disruptions to power connections (i.e. unrelated to widespread events), but even this is estimated to be an economic cost of $150 billion per year.

The urban myth—that after prolonged power blackouts the results of people being trapped together in elevators and trains, or without home entertainments, has been followed nine months later by a baby boom—may have an element of truth. My examples that hit very large populations are all for relatively short-term failures of electrical power (i.e. less than a day). However, the baby boom statistics may become more obvious for power losses that last a few weeks, although up until now these long-term failures have mostly involved small communities.

For the electrical grid systems, the exposed cable networks are extensive, with high-voltage power transmission lines on pylons. Viewed in other terms, these are extremely large and efficient antennae. Many have power transmission networks over distances approaching a thousand miles in a single link (i.e. a superbly long antenna). For interconnected grid arrangements, the antennae lengths are considerably more. Aurora events can therefore induce surprisingly large additional voltages and currents in such network lines. Although mains voltages are, say, 240 volts in Europe or 110 volts in North America, transmission losses over long distances are severe at these voltages. Therefore the power grids operate at voltages above one hundred thousand (100,000) volts, as power losses can be reduced at the higher values. Actual numbers are unimportant for this disaster plot, but some systems run up to 750 kilovolts. The UK grid typically has the higher-voltage sections operating at 275,000 or 400,000 volts. These numbers are so far from 240 volts that there is a lot of complexity needed to transform the power down to the voltages we use as consumers. Even further complexity is needed to maintain a constant frequency (e.g. 50 Hertz in the UK). The equipment that does all of this is critical and expensive, so if any section is destroyed, it instantly puts further pressure on power sent via alternative pathways. It is great technology, but it has the potential to collapse under conditions of extreme electrical overload.

Despite the high-quality engineering, the voltage and power surges caused by solar events can drive the power grids into an overload mode. Damage scenarios are many and well documented. Power cables can melt or arc down to the ground, which can cause pylons to collapse. Alternatively, the power surges rip through the transformer stations, and the extra power is sufficient to burn out both the transformers and the frequency control circuitry.

During this century, power cable systems have become much longer, more complex, and more interlinked, with the result that even small flares have produced blackouts and destroyed transformers and other equipment. In the last 15 years, there have been numerous examples, particularly in high-latitude countries such as Sweden and Canada. This is to be expected, as the aurora effects are most visible there, but really major storms can be seen much farther away from the poles, and induce far greater electrical disturbances. At high latitudes some precautionary measures exist, but at lower latitudes (where aurora type interference is rare) there may be no attempt at building adequate protection against major flares. For interconnected regions and countries, the system is endangered by the weakest link of this electrical supply chain.

Damage to a single transformer in an extended network can normally be handled, whilst repairs are being made, by rerouting power via alternative paths in the grid. If more than one section of the grid is closed, then alternative power routing may not be possible. This difficulty has occurred in the USA where, as a result of overload trips cutting in, there has been a total power blackout over large regions. The serious longer-term problems occur when more than one transformer unit is destroyed (as could happen in the solar flare scenario). Replacements are far more problematic, because the electricity companies do not keep enough spares (they are too varied and too expensive). Therefore loss of a single transformer may cause problems for many months. However, multiple losses would be catastrophic and long term, as they could cause a power loss over a large region. In extreme cases, it will be an entire country: there would not be manufacturing capability to build the replacements, and the country would be totally dependent on foreign suppliers. They would also be vulnerable to exploitation or invasion.

The real disaster situation

So my fantastic disaster movie scenario appears totally unexciting, as I am saying that just a reasonably large sunspot event is capable of destroying large sections of the electricity grid, and sunspot activity of an adequate destructive scale is not particularly rare. The only difference now, compared with earlier major sunspot events, is that we have immense electrical grid systems linked together that would act as antennae to feed in the destructive power. At local levels, there will of course be precisely the same types of damage with destruction of equipment and a spate of electrically driven fires (as for the 1921 and later events). There are enough previous examples of failure that the only discussion is about how often is it likely to happen on a large enough scale to have a real impact on the survival of a nation. Nevertheless, the possibility is so real that both Europe and America (because of their northern latitudes) have commissioned studies to see the probable scale of the catastrophe, to guess at the death tolls and the economic losses, as well as the time it would take for the nations to recover. The official reports make grim reading.

The 'official' US-based study is actually considerably more optimistic than some others. My guess is that this is because within the USA the most vulnerable region in terms of latitude and electrical conductivity of the ground (actually an important factor) means that only the eastern seaboard north of, say, New Jersey and New York are at major risk. The assumption is that supplies and assistance can be mobilized from the more southern and western states. Personally I think this is overly optimistic, as earlier claims of the 1960s—that in the event of a nuclear attack, the main facilities would all be running within eight hours—definitely did not match the reality of an electricity grid problem in the same period. Cities such as New York were powerless for well over 24 hours. That was a trivial event compared with the probable sunspot-driven electrical failures.

Using the 1859-size Carrington event as a benchmark and applying it to current grid supplies, I can quote an official US study as suggesting that the storm would have dramatic impact directly on 'between 20 to 40 million people with durations of 16 days to 1 or 2 years'! The economic downside from this is guessed at as being a few trillion dollars. Now reread this paragraph, and consider what these bland words mean to real people.

When will it happen?

I will remind us that nothing on the scale of the Carrington 1859 event (or anything larger) has hit the earth since then. Although an equally powerful storm in 2012 crossed Earth's orbit, it fortunately missed us by nine days. In solar terms, these events are not rare, or particularly large, and the records suggest that Earth is struck roughly once or so every 100 to 150 years with magnetic storms on this scale. Much bigger flares are known to have existed, as they left chemical signatures in the ice caps. The big ones may only be once every 500 years. These are semi-random events, so an average of, say, one every 100 years could be misleading, and we might have several in a much shorter space of time. Nevertheless, 1859 is already well over a century ago, so the probability of a new comparable bombardment is rising rather steeply; experts on this topic cite a 10 per cent chance that the next major one will happen within a few years, with a 100 per cent value by the end of the century.

The other unfortunate feature of the US study is that much of the data and estimates have not been made available to the public. This may mean that the information used in the analyses is considered to be classified, as it might be demonstrating weaknesses, or potential terrorist targets, or, equally, it may be that the data are not published to avoid fear and anxiety.

The very unwelcome conclusion is that a real disaster scenario is either imminent within our lifetime or very highly probable for the next generation. We need to look seriously at the potential consequences of our love affair with electronics and the danger of solar flares. The news is not good. Starting with the electrical power distribution lines, there is an immediate and urgent priority to ensure they will survive, even if there are closures for a few days. Relative to the money needed for repairs, this is small, but the concern is that for short-term economic reasons, no power companies will make adequate investments. There is the further concern that in some countries, power generation and distribution are run by different companies, and each will of course have shareholders who assume that the protection and the costs involved are the responsibility of the other group. The logical suggestion is that the funding should come from a governmental source, as the collapse of the power networks would hit all the country, both civilian and military.

US estimates of power failures lasting more than a few days, as have occurred in the last ten years in a number of countries, put the costs for each of them in billions of dollars from lost productivity and insurance claims. So far all the power loss events have been localized or minor, and mostly due to human error and equipment failure, not from natural events such as solar flare activity. Predictions on more significant power loss, for, say, a month, run into trillions of dollars, with very different guesses at the overall outcomes and death tolls. In my scenario of a truly major event equivalent to that of 1859, there would be a total collapse of a country such as the UK. Therefore, I believe that protective energy grid measures should be funded as a priority by central governments. A power grid collapse would be more disastrous than anything we have experienced in the past, and it would expose us to civil and military chaos on an unprecedented scale.

How bad could it be?

A key factor to add into any such disaster scenario is the state of the weather and time of year. For the eastern side of North America, the Canadian average temperatures for January in Montreal and Quebec City can be as low as −9°C and −13°C (16°F and 9°F), respectively, whereas New York is much warmer, at −3°C (27°F). In each case these temperatures are warmer than would occur for a city without power and transport. So power loss of even the minimum 16-day estimate would involve incredible hardship just in terms of heating and survival. The larger cities would be gridlocked as a result of no lights and traffic control, and people panicking and trying to escape from the city. Fires are inevitable (as for 1921, etc.), and dealing with them would be extremely difficult in a gridlocked city. Fire problems would be made far worse by lack of power to pump water, so many fires would become out of control.

Much of London was destroyed in the Great Fire of 1666, but we assume that this is less likely now, because the houses then were mostly wooden. However, the example of the Twin Towers atrocity shows that even modern skyscrapers can rapidly become an inferno. In any major city where a dozen such fires were triggered by voltage pulses from the aurora, there would be chaos, widespread destruction, and considerable loss of life.

Fire danger is not confined to cities and electricity substations or generators: the pylons or other cables strung across the countryside funnel the energy to any available grounding site. Canadian examples have shown that trees growing towards the pylon wires act as excellent paths to short out the excess charges and, just as from a lightning strike, are major sources of forest fires. Therefore both forests and farmland countryside, or even urban gardens, may be potential fire hazards under the extreme aurora conditions.

City dwellers would be at an incredible risk, and the casualty level would be high. This is a direct consequence of their total reliance on technology for power and survival. (More remote areas would be more likely to have facilities that survive.) Health risks would be dramatic without access to medical supplies and treatment. Isolation in tall tower blocks, and failure of water, electricity, gas, sewage, and non-functioning shops or food supplies would mean that after a few days, most people would be running low on water and food; hygiene would be non-existent. Without food, we can survive for many days, but without water, life expectancy is probably far less than a week.

Wider area consequences of grid failures

I have focussed on the major damage sites, but such a storm would bring minor failures over a much wider region, even at lower latitudes, as the telephone and power lines in many cases are running over long distances. So they will be precisely as vulnerable as the 1859 telegraph system. Home devices certainly would not survive, and central industrial units are likely to have the same fire problems as in 1859 and 1921.

Other electricity-generating sites that did not exist in former times are arrays of solar panels, wind farms, and nuclear reactors. Each type tends to be far from the city users, so are connected by long cable networks that are able to couple in the excess power. This in principle could destroy the various units. In the case of nuclear systems, a truly unfortunate combination of events might remove the power that is needed for an automatic shutdown of a reactor. As seen from Russian, Japanese, and American examples, inadequate design, poor materials, and unpredicted natural events can lead to severe problems, which would contribute to the overall disaster scenario. Currently there are changes being made in some types of reactor because it was realized that they could not cope with a simultaneous loss of local and external grid power.

In country districts, farming is still dependent on fuel and electrical power, and with ever-increasing intensive farming and larger fields, their failure would impact food supplies in the immediate short term, and the ability to plant and harvest in the longer term. Details would depend on the timing of the event during the farming year.

Given the prediction that it could take months to manufacture and install replacement power grid equipment, the likely, and necessary, result is that city dwellers would try to move to the country. In a large country such as the USA, this could be organized, although the chaos from the event would be enormous, and on the basis of up to 40 million people affected by such a solar event, it is inevitable that there would be major lawlessness, and the introduction of martial law.

Which areas of the globe are at risk?

If I assume that New York is at the lower edge of the latitudes seriously hit by the storm, then in European terms this means everywhere north of Madrid will be equally vulnerable. The pattern of damage, chaos, social collapse, and effectively national bankruptcy will prevail across the whole of Europe. Although there have been solar storm–induced power failures in South Africa, I have taken the hopeful view that if the intensity is only critical to latitudes higher than 40° (i.e. north of New York, Madrid, or Beijing), then virtually all countries in the southern hemisphere would be exposed to much lower total damage, as there are fewer major cities. If the sunspot activity were more intense, then electrical storms would of course cause disasters at lower latitudes in both hemispheres.

Overall, our vulnerability to a sudden major collapse of advanced civilization is the potential price of dependence on high-level technology. Without communications, food distribution, and means of sustaining government, the dangers are initially from within. People will sink to a level of personal survival, which may well be highly selfish. The external factors are equally unpalatable, as the European countries plunged into chaos will be unable to mount strong defences against invasions from expansionist military neighbours who have survived. We will equally be obvious targets for religious crusades by extremist fundamentalists. If we consider what happened to Central America with the expansionist/religious invasions in the fifteenth century, then the prospects for current European generations are remarkably bleak.

Is there any good news?

Thinking about the potential for collapse of many advanced countries is extremely depressing, but there clearly is a positive side in terms of humanity, as the chaos will not impact totally across the entire world. There will certainly be regions that will survive in underdeveloped nations, and pockets of survival of independent communities in more remote areas of the larger continents. Surprisingly, I think the answer to my question is a qualified 'Yes'. There is no question that the pattern of world trade and technology will be irrevocably altered, and the total death toll would be measured in many tens of millions. On the other hand, this is a small percentage of the total world population. History tells us that empires rise and fall, and effectively this disaster would just precipitate an equivalent shift in the balance of power and trade across the globe. Therefore, my positive view is that the human race will survive, but power, ideology, manufacture, and trade will be fought over by nations who will try to exploit the gap left by the temporary demise of industrialized northern nations.

Again I am being positive by saying 'temporary'. Nations have survived wars and plagues in the past, and rebuilt. The example of the Black Death in Europe, which killed a third of the population, produced economic and social chaos, and caused a total restructuring of the social order. However, it can be viewed as a positive precedent, as European civilization eventually recovered, albeit with changes in the social structure. We were fortunate that the plague faded away, whereas other diseases have emerged and remained for many generations. Similarly, in the two major wars of the twentieth century, death tolls were also in the tens of millions, and after the 1914–18 conflict it was compounded by an influenza epidemic that may have killed as many as a further 30 million. Yet the countries involved have recovered.

Loss of a particular civilization has happened for many reasons, either natural events or wars, invasions, religious crusades, and 'conversions'. With each disaster, humans have lost knowledge, skills, and culture of that region and their preceding generations. The information and cultural losses are more important than individual people, who have a limited lifespan, so it is particularly sad that by far the greatest damage and loss of life has invariably been self-inflicted, typically as a result of political or religious warfare, and we have used technology to produce ever more destructive weaponry. Despite this, the human race is surviving and evolving.

Humans are clearly a resilient species, but there is no guarantee that we will ever learn lessons from our mistakes.

Knowledge is power and absolutely essential for survival

In this chapter, I have just considered a modest-scale natural event, with potentially disastrous outcomes for a large number of people on the planet precisely because we are dependent on advanced technology. The catastrophe imagined here is not just feasible, but statistically highly probable. My concern is that the predictions of the ensuing chaos may be too modest. The root cause of solar activity is not unusual, even in recorded history. The only difference is now we are vulnerable because of electronics.

We have intelligence, determination, knowledge, and the ability to plan for the future. In my disaster models, the potential dangers from a simple natural occurrence are very clear, and there is no reason we cannot put all our skills and knowledge to a positive purpose to prepare ourselves for them. It will take money and effort, but the alternatives are so serious that we must be prepared. It may not have the popular appeal of sending people into space, but the preparations for non-satellite communication, or buried cable replacements and safety features on electrical grid systems, are far more important. Such activity may be happening in the background, as for the two satellites looking for a storm of solar particles, but faster circuit breakers, spare grid transformers, and ground-based optical fibre communications, etc. need international support, not ad hoc planning. With it we can survive; without we will not.

How should we view technology?

For me the key lesson is that, whilst technologies bring many benefits, we become dependent on them and invariably are highly blinkered in recognizing the obvious downsides of technical progress, and are slow to react to the unexpected negative features of our ingenuity. Therefore in the remainder of this book, my intention is to highlight examples of technology that seem desirable, and show that many have dark sides that are not in our best interests. Some problems have arisen from ignorance, others from greed and a desire for instant profit, plus an apparently inherent desire to exploit others for personal or corporate gain. This is mirrored in our aggressive actions of warfare and physical or economic enslavement of others. The pattern is not new. Stone axes made fine weapons as well as tools for survival. The only difference between the scale and destruction of historical conflicts and modern ones is that weaponry and propaganda have greatly advanced. Indeed our technological progress has very often been funded and developed with military aims in mind; subsequent benefits were just by-products. As a trivial example, few people realize that something as familiar as photochromic sunglasses, which darken in the sun, are a by-product of the result of attempts to make glasses to stop soldiers being blinded by the flash of nuclear explosions.

From a positive perspective, technological advances have brought innumerable improvements to our lives; therefore we need understanding and ever deeper insights into how we respond socially, as well as in the details of the various sciences and production processes.

On the basis that knowledge is power, the assumption is that we should be eager to learn new ideas and processes, and scientific advances will always be welcomed and beneficial. This is false on each count. Quite surprisingly, we have a great reluctance to learn new ideas, preferring the security of familiar ones (even when they are wrong). Similarly, many technologies are developing for commercially driven reasons, not for the customers. Certainly, new ideas are needed if existing techniques are inadequate, for example, to increase the capacity of Internet connections, as the system is continuously battling with a usage demand growing faster than the technologies can sustain. Here the options are very limited, and reduced usage may be enforced by pricing or other restrictive practices.

Many others 'advances' are non-essential and intended to boost sales of a product. Often the replacement items are incompatible with existing systems, which then become obsolete. Computer and mobile phone systems immediately provide such examples. Not only is this unfortunate, as the cumulative effects can mean a great deal of information loss, but it also results in losses of former skills. Less often considered by the industries is that the changes can isolate and undermine large sections of the population, particularly the elderly and less affluent.

There are many variants of such losses; I plan to outline a few of the reasons for them and the driving factors that deprived us of accumulated skills and knowledge. Far less dramatic, but equally effective, are the mechanisms that gradually erode knowledge and history because we have advanced in the technologies of our lifestyles or modified our religious or political ideas. There are a multitude of examples, ranging from the loss of skills in making flint tools (because we no longer need them) to loss of records, documents, and photographs because computer upgrades have made our files and storage systems obsolete.

One clear trend emerges throughout this book: the faster we make technological advances, the shorter the survival time of records and information. In an era when we glorify rapid technological progress and embrace it as soon as possible, we rarely realize that this is condemning our stored information to rapid oblivion. We may have photos of Victorian ancestors, even their love letters, but the present trend is that our current digital images and correspondence will probably be unreadable in 20 years' time, and certainly will be lost to our grandchildren and future historians.

Electronics will appear in various guises, as it impacts on many technologies that are rapidly advancing on an unprecedented scale. Within the lifetime of many, we have gone from virtually no TV just 60 years ago, no home-based powerful computers until 30 years ago, and no mobile phones until this century. The changes have happened so fast that we are completely overwhelmed and impressed. We want more, and are moving so rapidly that we never consider what would happen if we were forced back into the technological world of our grandparents (or earlier!).

Losses of past knowledge encompass language, culture, literature, art, and music, as well as the tangible decay and destruction of documents. The field is so wide that my choice of examples is quite personal, but they all fit the same overall pattern. Although harder to quantify, but equally important, are the ways in which new technologies are isolating and separating different blocks of society and replacing human contacts with merely electronic ones. A 2015 survey reveals that 14-year-olds in the UK spend eight hours a day looking at computer or mobile phone screens (and there are more short-sighted young people than before). This change in interactive behaviour has repercussions that are extremely difficult to predict. More obvious as immediate losers are the old and the poor, who cannot cope or cannot afford new equipment. They suffer increasing isolation from the technological world. Such problems are overlooked by the inherent arrogance and optimism of the young, driven by commercial and governments interests. By the time the consequences are appreciated, the effects may be irreversible.

The following chapters explore some of these ideas and the consequences of technological advances, together with the dark side we fail to see in our enthusiasm from the marketing of gadgets, new toys, 'must-have' electronics, and new software forced upon us by 'upgrades'.

3

Natural Disasters and Civilization

A fascination with danger

Disasters and death have always been an integral part of our lives, and we have a morbid fascination with both. For entertainment, we enjoy storytelling, books, films, and TV, especially where the plots involve possible misfortunes, magic, supernatural powers, aliens, murder, and death. I imagine psychologists have many theories for this, but basically it seems we only have a tenuous grip on life and so enjoy hearing of dangers that will scare us, but enjoy them because we know they are just fiction. Far fewer people actively take pleasure from reading of real-life tragedies and suffering, although we can be deeply moved by such stories and histories. Nevertheless, reality should motivate us to act, so there are times when we must face up to potential misfortunes, especially if action may help us to survive them. Therefore in our fascination with natural disasters, it is worth trying to separate those events that are inevitable and beyond our control from those that could be mitigated by thoughtful preparation. All too often we fail make any effort. Typically, we say that the preparations and looking at problems are too difficult for us; the effort should be made by the council or government; or, equally frequently, we take a lazy option and say we have no control over our lives, or all events are an act of God. Religion is then an easy way to avoid personal responsibility.

My opening chapter offered an example of natural, frequently occurring events, which were once of minor importance to our survival, but have now been placed in a very different role because of our dramatic shift to a technologically based society for the 'advanced' nations. Interestingly, Third World lifestyles will be far less harmed by the loss of satellites, power grids, and collapse of very large cities. Perhaps 'the meek shall inherit the earth' has some truth.

This is actually a very basic and fairly standard forecast for a large range of serious events, both natural and constructed, because for the Third

World, a major catastrophe that undermined the technologically based nations would allow them to move into the vacant niche that would appear. The later chapters will concentrate on the negative aspects of a vast number of improvements in all aspects of our lives from agriculture, communications, medicine, transport, technologies, and warfare. In most cases, we are able to see there are clear advantages for us (including warfare, if we are the winners), but I emphasize that we invariably look at short-term gains and ignore long-term dangers. The negative aspects may well be global, but without exception they are likely to be most critical and apparent for the advanced societies. They are endangered most, not least because the 'advanced' nations rely heavily on technologies that are unlikely to survive globally damaging events.

This is not pessimism, as inevitably anticipation and forward planning will be valuable, as they will help us reduce the impact of various natural catastrophes. Of particular importance for humanity is that, unless the damage event is global, we need to make every effort to preserve our knowledge, skills, and information in forms that are accessible and distributed across the world. The obvious example of this defensive strategy is not to have all information stored only on an electronic system that is solely accessible by electronic communication via satellite, or on computers that need electrical power. Chapter 1 showed that for modest natural events, that are not exceptionally rare, there could be an irreversible destruction of knowledge and a loss of data, even if the mortality rate were low. At the personal level, most of us have already experienced the fact that, whilst our old electronic files and documents may exist, they are in formats and storage systems that make them inaccessible. Not only have I had precisely this problem, but as a cautious man, I have kept back-up files. Despite this wisdom, one computer had the back-up drive placed physically above the main drive. One of them jammed, overheated, and destroyed the other! (I now use separate systems for back-up.)

The dangers of advancing technologies are thus immediately apparent for all of us, and future progress in computing and electronic storage formats will make the problem worse, not better.

Many types of natural disaster have occurred in the past, and are likely to recur in the future. They are similar in that at an individual level they are outside our control, but all will devastate our lives. Natural disasters are intractable as we have no way to predict or avoid them, and must accept that they are part of the pleasure of living on this planet. For the truly global catastrophes, our hope is that some humans will survive; if not, different life forms will replace us. In planetary terms, we have not existed for very long, and our rapid rise may be matched by an equally rapid decline and exit. Only in our imagination are we the inheritors of the earth and immortal as a species. Philosophical dinosaurs probably also assumed they were the ultimate species.

Events that happen only on geological timescales

In geological history there have been many extinction events, and in each case the majority of species have been annihilated, but so far there have always been a few life forms that have survived. Genetic diversity is impressive and highly varied. This is great news, as it allows the survivors to evolve to fill the ecological niches that followed the extinction event. The most familiar and frequently cited example is the meteor impact in the Yucatan peninsula. It has gained a widespread press because some 65 million years ago it probably caused, or contributed to, the demise of the dinosaurs (and nearly everything else). A word of caution is that in that same era there was ongoing intense volcanic activity in the Indian subcontinent, which was also a major contributor to a mass extinction. The physical size of the dinosaurs was impressive, and we have bones and skeletons that have captured our imagination. Especially for children, the dinosaurs fit the image of fairy-tale dragons, and so are really exciting. However, their disappearance was not the first massive extinction event, and certainly will not be the last.

Meteor impacts have happened regularly throughout the history of the earth; so far at least ten major craters have been found from meteors and other bodies crashing into our surface (others craters may have been covered over or vanished). In terms of diameter, the top ten largest ones we know about have diameters of at least 200 kilometres (~120 miles) and happened on timescales from 2 billion to 35 million years ago. Many others look less impressive but include far more recent events. In fact, there is absolutely no reason we may not have large-scale events in the future. It is just that the frequency is so low that we assume it will never happen in our lifetime. Our catalogue of the top ten major impacts will have missed out many equally large collisions with meteors that crashed into the ocean. Earth's surface is roughly three quarters covered with water so, at a stroke, I am claiming the number of big hits should be multiplied by 4. To me it seems odd that others do not seem to emphasize this. Perhaps the blast of water into the atmosphere, rather than rock and debris, has less damaging effects, although there would be immense tsunami waves hitting coastal regions.

In all cases, land or sea, the problem with a meteor that happens to bump into us is that it need not be very large, but because the impact speeds will be immense, so will the kinetic energy. So the famous Yucatan impact has a crater area of around 5,400 square kilometres. For me, this is meaningless, as I have no grasp of the scale, but in recognizable terms this is around the area of Wales (in Europe) or New Hampshire or New Jersey (in the USA). Now I am impressed.

Many other large meteor impacts have taken place, and small meteors are quite common. Indeed, they may even have been the source of water on the planet. The current problem, even for a medium-size impact, is not from the damage occurring at the initial collision site, but from the debris thrown into the upper atmosphere. This ejected material will circulate the world for some time (months or years), as once it reaches the stratosphere it is not washed down by clouds and rain. The obvious effect is to block sunlight, and so cause cooling of the surface, and this in turn means crop failures that destroy harvests and agriculture for many years. This is exactly the same problem that arose from major volcanic eruptions that blocked sunlight, lowered the temperature, and caused famine. One such example was the Icelandic eruptions in the 1780s that contributed to European crop failures for several years. Additional problems at that time were caused by shifts in the Pacific Ocean current, the El Niño, which brought drought to South America and a European summer drought in 1787, followed by excessive spring rains. These all contributed to the food shortages that probably catalyzed the French Revolution.

Climate-driven droughts—either from meteors, volcanoes, or changes in the ocean currents—have all resulted in famines, leading to loss of several civilizations or empires, and have been the underlying causes of revolutions. Examples range from droughts that are thought to have destroyed both the Aztecs and Incas in the Americas, led to the collapse of empires in Cambodia, and caused desertification of the Sahara, plus contributing to social collapse in many other regions.

With our twenty-first century fixation on electronics, and both satellite- and ground-based communications, the stratospheric debris will be considerably more devastating. It will obliterate much or all of our electronic communication systems. So, in this sense, as with the solar flare scenario, we will be in a far weaker position to survive in our present lifestyle than our ancestors. Yet again I am not predicting some dramatic doomsday scenario—only that without preparation and forethought, we are exposing ourselves to major communication and information loss, and severe consequences for our continued existence. Our hope is that survivors may have knowledge from simpler technologies that they could exploit. This is currently feasible, as we have written records but, as we move to totally electronic storage and dependence on other high-technology gadgets, then even this opportunity will be lost.

A very positive view is that in the past humans have not been immune to global catastrophes, and it is thought that early hominid populations plummeted on more than one occasion as a result of natural disasters; clearly they recovered and expanded. So my optimistic guess is that we would recover, even if at a very different level of civilization. We may be physically weak, but we are a resilient species, and at least we do not have to contend with dinosaurs who might wish to eat us.

Earthquakes and volcanoes

Viewed from a safe distance, and watched on TV, the spectacular and devastating natural events such as volcanic eruptions, earthquakes, tsunamis, and floods are awesome and exciting. Indeed they make great TV and news items, and people will willingly watch the repeats. Being there is of course another matter. The media presentations of such events are equally skewed by our inability to comprehend, or be deeply interested in, events that are far from home. An earthquake that kills a thousand people on the far side of the globe may, or may not, be reported on the inside pages of our local city paper. However, if two of the local community were injured in the event, then it will be front-page news. This is not a lack of humanity on our part, but a useful mechanism to cope with news items that do not immediately impact directly on us.

We live on a very dynamic planet so all such natural events are extremely frequent. In terms of location, both earthquakes and volcanoes tend to be primarily confined to tectonic plate boundaries. The plates are the blocks of land masses which slowly drift across the surface of the earth. Sometimes they slide past one another and cause earthquakes and very obvious fractures. These fractures between blocks can be extremely sharply defined; I have seen examples in California where there are side steps in field boundary fences! In the UK, a classic example is Loch Ness, which is bordered by two different plates drifting past each other. Looking at the rock structures on each side, it is obvious there has been a slide of some seven miles between the two banks.

Whilst we only see media reports of major events, with sensitive equipment, geological surveys detect more than a million earthquakes a year. So the level of activity is greater than most of us realize.

Alternatively, plates can hit head on, and then one is forced down underneath the other. Not only does this result in earthquakes, but it is a very favourable condition to allow volcanic activity. The entire rim of the Pacific is highly active geologically because of these plate collisions. The impacts are slow, but they involve immense quantities of energy, as seen from the spectacles of volcanoes and earthquakes. Effects are not just localized at the collision sites: volcanic eruptions eject massive quantities of rocks and dust into the upper atmosphere that circulate the globe for many years. Earthquake geology is not an exact science, but in the last 50 years we have gained knowledge about plate tectonics, and so now have a reasonable understanding of where such events are likely to occur. The very fact that they are at plate boundaries means they are likely to be in coastal areas (e.g. the Pacific Rim), and therefore are associated with ports and fertile regions that are heavily populated. Inevitably we want to farm in fertile regions, and if the good soil was created by volcanic activity, then we choose to assume the volcano will never kill us, despite the numerous historic examples of large populations being obliterated. Future geological events will certainly cause local extinctions, and with increasing human populations more people will be killed.

TV-worthy earthquakes happen perhaps 30 to 50 times a year, and the larger ones are often linked to volcanic activity. Volcanoes are visible spectacles; again we see around 50 or so a year. Unlike earthquakes, they are not sudden, short-lived events, and often continue for considerable periods. Therefore as many as 30 active systems are happening at any one time. Earthquakes vary in intensity; surprisingly, the majority are not obvious to us, as they do not make the ground shake under our feet.

In recorded history, the most devastating earthquake was in the Shaanxi province of China in 1556 when nearly a million people were killed (at a time when the world population was much smaller than at present). In the last 1,000 years, there have been a dozen other major earthquakes around the world with death tolls above 100,000, so our concern and fascination with them seems justified (especially from the relative safety of a region such as the UK, where major earthquakes are currently unlikely).

A spectacular volcanic event took place in the Mediterranean during the Bronze Age, which totally destroyed the Minoan civilization with the eruption of the volcanic island of Santorini. The explosion occurred around 1600 BC; it is the largest documented volcanic event in our history. From the human perspective, it was a particularly unfortunate eruption as there are traces of written records and paintings suggesting that the Minoans were a remarkably civilized and egalitarian society (not least in the apparent importance given to women). With such a large explosion, the effects of that particular volcano were not limited to the Mediterranean. Chinese records imply that the same event precipitated the demise of the Xia dynasty. Their political collapse was triggered by atmospheric debris that caused the climatic impacts of a yellow fog, summer frost, and famine for several years. The Chinese dating would place this near 1618 BC. The explosion may also have been the origin of the myth of the world of Atlantis, since Santorini dropped below sea level (except for the part that shot into the atmosphere).

In the spirit of the TV-type descriptions, I probably should quote how much material was thrown into the atmosphere. The island of Santorini was originally a modest-sized mountain that has been replaced by a residual sea-filled crater that is some 5 × 7 miles across. However, this crater volume only indicates how much island vanished, and we can only guess at the volume of magma that was also spewed out during the eruption. Some estimates suggest it was some 60 or more cubic miles of material. Imagining this amount is hard, but nevertheless it was 'a lot' (the precision of my scientific background is showing).

Genuine numbers do exist for other volcanic events. Within living history, the 1980 eruption of Mount St Helens in upper Washington state in the USA is estimated to have tossed out about 0.7 cubic miles of material. This was a modest amount and certainly not in the same league as the ejections from the Yellowstone volcanoes. The 45-mile-diameter Yellowstone National Park is in fact a mega-sized volcanic crater from which there have been regular eruptions on a massive scale. One of the largest of these blackened the atmosphere and left immense surface deposits across vast tracts of land, as it expelled some 2,500 times the material seen from Mount St Helens. This Yellowstone volcano was so phenomenal that it is hard to comprehend, but as a guide this is roughly equivalent to a column of material the height of Mont Blanc that would fill the area around London bounded by the M25 motorway. An equivalent estimate for reference to the USA would be Long Island buried to the height of Mount Washington.

Eventually, Yellowstone will offer a repeat performance; current estimates suggest the underlying magma chamber has spread out over some 20 × 50 miles. In terms of fallout of ash when it erupts, one can predict that the middle of the USA will be devastated, and even the eastern seaboard will be buried under centimetres of ash. So across the entire country, there will be certain death to exposed animals, total loss of crops and communications, and a human toll measured in many tens of millions.

We think of mega-scale volcanic explosions as only occurring rarely, on geological timescales of roughly 100,000 years. The big events have been spread at various locations around the world. By contrast, we do not think in terms of humans over such timescales. Nevertheless, we have found human cave paintings dating back 30,000 years, and hominid bones that date to even earlier. Therefore these two timescales are closer than we would normally consider.

A final slightly worrying fact is that the last three eruptions of Yellowstone were around 2.1, 1.3, and 0.64 million years ago. I plotted a graph of these events to try to estimate when the next one might take place (assuming there is an underlying pattern). The graph is remarkably smooth and it suggests the next eruption is overdue by maybe 20,000 years! Perhaps we do not need to worry, but in geological timescales it is imminent.

A second well-documented Mediterranean example was the eruption of Vesuvius in 79 AD. This was on a very much smaller scale, but it buried Pompeii and Herculaneum. Since I am mostly discussing information loss, we could view that eruption more positively, as for archaeologists it offered a buried time capsule of life from the period. An observer at the time, Pliny the Younger, gave a detailed account of the eruption, including a description of ashes moving like a flood at high speed down the mountainside. His report of high-speed flow was originally discredited, but the same pattern was recorded in 1980 from the Mount St Helens eruption. This flood of solid and expanding hot gases is now termed a pyroclastic flow and examples have been seen to move at extremely high speeds.

Vesuvius is still an active volcano that poses a threat to Naples. Perhaps less obvious is that it is a small part of a much larger system, and the entire Bay of Naples is actually a crater that is the remnant of a massive volcano that erupted at some earlier period.

No summary of major volcanic events would be complete in the popular press without mentioning the major 1883 event when the island of Krakatoa in Indonesia erupted with massive quantities of material flung upwards in a series of explosions over the course of several weeks. The known death toll was around 36,000 people from ash and related tsunamis. The main explosion was so loud that it is claimed the sonic boom was heard in Europe. The vast dust and ash cloud modified global temperatures and produced colourful sunsets for several years after the event. In fact, Krakatoa is part of a set of volcanic islands around a central caldera that has erupted many times previously, and indeed the site is growing again with a new central cone that will erupt again.

In the spirit of capitalizing on our penchant for disasters, a film was made in 1969 that used elements of the actual eruption as the basis for the plot. The film went under the name *Krakatoa, East of Java*. The film has a nice emotive ring to the title, but unfortunately Krakatoa is actually west of Java, in the strait between Java and Sumatra!

Future eruptions

More eruptions and earthquakes will follow in sites of high population densities such as Japan, Turkey, and California, as all of them are sitting on highly active tectonic plate boundaries. From their location there will inevitably continue to be serious consequences with each future earth movement. Prediction is difficult, but earthquakes cause stress relief in one region that moves the next likely site farther along the stress line. So, for example, there has been a steady and roughly predictable pattern of earthquakes moving westward along the top of Turkey. The sites have been estimated rather more reliably than the timescale. Unfortunately, it implies that cities such as Izmir and Istanbul (which was once the largest city in the Western world) will eventually be hit by major earthquakes at some point within very modest geological (and human) timescales.

With a global economy, the influence of these disasters can have surprising long-term, and long-distance, repercussions. I remember trying to buy parts for a laser printer, but these were only made in Kobe, Japan, and a major earthquake there caused a hiccup in the world supply. There are many other unique factories whose loss would have similar long-range effects. This type of problem is one of the consequences of technological progress. Rather than being dependent only on local food supplies and resources, we are now using distant suppliers and key minerals and ores that can be highly specialized in terms of origin. Loss of the resources—whether due to natural or human-induced disasters—will pose severe difficulties. We are equally vulnerable to political control of key items.

In general, earthquakes are locally devastating, but the rest of the world will survive. The only caveats are that the economic impacts caused by devastation of large industrial areas are felt for many years after the event. So, for example, a Californian quake could destroy Silicon Valley, which would be a major economic problem for the USA, but globally the skills would be retained.

Equally, China, an increasingly major industrial power, covers a very large land area, and has regularly had severe earthquakes. Inevitably these will sometimes destroy major cities and industrial complexes. A particularly bad scenario would be the loss the major hydroelectric power station on the Yangtze River, the Three Gorges Dam. This is the largest hydroelectric system ever built, with a power generating capacity of 22,500 megawatts. Damage or collapse would not merely result in a high death toll, but also inhibit industry and function of a large section of the country. Earthquakes and flooding in this case may not only be from natural events, as the change in water level involved from the construction has resulted in soil instability; nearly 100 landslides have been reported since completion in 2010. A major collapse is highly unlikely, but the sheer scale and novelty of the construction means the design is at the forefront of engineering technology. It has already reduced the incidence of natural floods (i.e. a positive feature), but unexpected and unfavourable long-term effects may appear for any engineering project on this scale.

Of the inevitable natural disasters, among the most serious and probable global events are likely to be caused by medium-size volcanic eruptions, not least as there are many of them and eruptions are happening every day somewhere in the world. The Icelandic examples earlier this century were not of a particularly large scale and continued only for a few weeks, but the corrosive dust particles were a serious hazard for planes. By contrast, when the next Yellowstone eruption occurs, North America, as we currently know it, will end. The long-term effects are very hard to predict, except that it is sure the entire world civilization will change and economic and military power structures will alter dramatically across the globe. The reasons for this include the circulating ash cloud that will cause crop failures and famine (particularly in the northern hemisphere), economic collapse of many countries, and loss of two or more major world powers.

European effects from Icelandic volcanoes

Because for the UK, the Icelandic eruptions are the ones most likely to cause a local effect, I will briefly review events that have produced European problems in recent times. Iceland is an active region with at least 30 separate volcanoes. A major eruption event in 1783 from the one named Laki produced a thick fog and ash over Europe, causing cooling, crop failure, and famine; as already mentioned, it also triggered the unrest leading to the French Revolution.

A different Icelandic volcano, Eyjafjallajokull (besides being unpronounceable in English, and therefore rarely discussed verbally), erupted in 2010 and put out fine ash that closed North European air traffic. The caution in blocking air traffic was due to the danger from the fine ash that is produced when volcanic ejections interact with water and ice, as from the Icelandic glaciers. The water and steam fragment the large lumps of ejected material into very fine particles of ash. This dust is then supported for a long time in the atmosphere, often at the level of air traffic. Because it is so fine, it is very difficult to detect, it is not as visible as a cloud of smoke. When the ash is sucked into jet engines, it is melted into glass droplets in the intense flame zone (which approaches 2000°C). However, as the droplets leave the engine, they cool down to, say, 1000°C, then stick to the engine components in the form of glass. There have been previous dramatic examples where the buildup of glass has resulted in jet engine failure and plane crashes.

Since the atmospheric magma particles are finely dispersed, they are not easy to detect, so it is tempting to assume the problem is negligible. Nevertheless, a modern jet can push some 60,000 kg of air through the engine each hour. (Because planes are heavy, they need a lot of energy to keep going.) So even if the dust is only at very modest concentrations of one part per thousand of the mass of the air, and just one per cent of it sticks on the engine surfaces, then the glass buildup would be at a rate of kilograms per hour (i.e. guaranteed engine disaster). Interestingly, this is an example of a problem caused by improved technology, as in earlier aircraft the internal engine temperatures were not high enough to form glass droplets. In 2010, closing the North European air traffic may have appeared to be a non-event, but it cost the global economy some $5 billion. Nevertheless, the costs are far less than the indirect effects of air crashes.

Air control policy has altered (possibly influenced by the economic impact), and it is assumed that there may (may!) be less of a hazard than was assumed in 2010. However, if there is a repeat eruption, my guess is that I will be among an increased number of Eurostar train users.

Another interesting Icelandic volcano is Bárðarbunga (also hard to pronounce), which is beneath an ice cap. The cap is steadily melting due to a significant rise in temperature at these latitudes. Because the surface is lifting at one foot per year in some regions, glacier weight is reduced over the cap of the volcano. Inevitable consequences are more eruptions and local flooding, and of course the potential to have fine ash ejection into the atmosphere. Continued monitoring is in progress, not least as Icelandic volcanic activity has rapidly increased within recorded human history by some 30 times.

Tsunamis and floods

Invariably tsunamis and floods are localized problems; quite unpredictable and locally devastating, but unlikely to alter the course of the world in general. Tsunamis are not only caused by undersea earthquakes, but also by underwater landslides. Therefore they need not be confined to tectonic plate boundary regions. For example, a Norwegian fjord landslide a few thousand years ago sent a wave hurtling across the North Sea that carried sand and debris some 50 miles across the flat fenlands of England. An analogy of such events is water sloshing along a bath when something is dropped in at one end. More major examples are underwater landslides on the volcanoes of Hawaii sending a surge that rose up mountainsides in North Australia.

Several TV programmes have capitalized on our interest in these natural events by searching for truly mega events that might possibly occur. We watch them—we are impressed and enjoy the programmes—but deep down, we assume that they will never happen. Nevertheless, it should be noted that there is a genuine possibility (indeed a certainty at some future date) that there will be a landslide in the Cape Verde islands that will send a mega tsunami across the Atlantic, destroying the cities on the east coast of the USA. The scale of the event is certainly 'mega' as the wave will be several hundred feet high on arrival and persist for hours.

The energy involved, and the prolonged duration of the tsunami wave, will totally destroy the eastern seaboard of the USA. It will kill a vast number of people in the cities, and economically will bring chaos to the countries that it strikes. Tsunamis may not look impressive as they race across a deep ocean—their destructive effects emerge only as they rise up when they hit shallow water. They can travel at speeds up to 600 miles per hour, so advance warnings of a few hours would be possible, but still inadequate in terms of allowing many people to evacuate.

Eastern USA is in line with the predicted main wave front, but the sideways effects will send lesser waves up and down the Atlantic. Furthermore, the shape of the English Channel will cause the wave to be forced ever higher as it reaches the narrow strait between Dover and Calais. For the coastal towns on both sides of this waterway, there will be immense damage and flooding. (Before panicking and relocating, relax: it will happen, but it may be many millennia before it does.)

Rain storms

Technology and meteorology have improved our understanding of causes of floods; it is now known that the upper-atmosphere jet streams can contain 'rivers' of rain. ('River' is suitably emotive for the media, but meteorologically is inaccurate.) A moderately recent example of a severe flood caused by such an event happened in 1862. It was driven by a period of prolonged intense rain for around 40 days from a northern jet stream 'river' that had drifted unusually far south. This inundated Sacramento and the associated valleys of California. It similarly severely affected all the neighbouring states. For example, it formed a lake in Arizona 60 miles long and 30 miles wide. If the weather patterns duplicate this flood in our time, the effect would be even greater, because the Sacramento Valley floor has sunk over the last 200 years. The flood would therefore be much deeper. Once again, I can point to the role of technology leading to an increased danger, as the sinking of the ground level is the result of arterial water extraction for agriculture. In 1862, the event caused economic ruin across the entire region; we can expect the same financial impact from future floods caused by this type of source.

My examples so far are for relatively short-term natural events, with long-term consequences for us. I have already mentioned that droughts were devastating and are blamed for the loss of civilizations from the Americas to Asia. But less sudden changes in weather patterns can be equally destructive and long term. A minor change in average temperature disrupted a monsoon weather pattern that had delivered rainfall and fertilized North Africa until about 5,000 years ago. A rapid switch from fertile savannah to the sands of the Sahara Desert emphasizes how delicately balanced is our reliance on climate. (The term 'Sahara Desert' is overkill, as 'Sahara' means 'desert'.)

We contribute to sensitivity to such changes by farming practices and destruction of natural forests, because deforestation can reduce rainfall. Once lost, forests may not recover. The scenario of an equivalent desertification of the South American jungles initially seems ridiculous, but the parallel with the sudden changes that happened to the Sahara is unfortunately much closer in time, and more probable, than we might expect.

Ice ages

Whilst considering past, present, and potential future natural 'disasters', I should at least include a mention of ice and ice ages. They have been a frequent part of Earth's climatic history, although, fortunately, we are currently enjoying a warm interglacial period. Their recurrent pattern, as measured from the ice core records from earlier ice ages, suggests they are partially predictable. Ice ages occur for a number of reasons, including small fluctuations in solar activity, and minor deviations in the orbit of the earth that have taken us slightly farther from the sun. We know the scale of these periodic orbital variations; we also know that the angle at which Earth is tilted relative to the axis of rotation steadily changes in a fixed pattern (termed precession).

Precession is not a new discovery—it was appreciated from the early astronomical observations of a few thousand years ago. It was detected because the direction to the Pole Star is not quite constant. The precession is quite slow, so this influence on the climate is of no immediate concern. However, the timescale before the next ice age occurs is difficult to predict because of the number of minor orbital oddities that need to combine to cause it. The estimates therefore vary between a pessimistic one—with another ice age happening in 2,000 years—to a more optimistic estimate of 30,000 years. No matter which is correct, for our generation, it is not a pressing problem.

Instead, we should be more concerned about survival in a world that is steadily heating. Confusingly, with a complex situation of many inputs to the earth's climate, some models suggest that heating contributes to ice melting, which may in turn influence a shift in the rotation axis and tilt, leading to a more imminent ice age. No matter which models are correct, glacial events will certainly reoccur. If civilization is well organized, we should survive, but with a greatly reduced sustainable population, ideally achieved intentionally rather than by war and chaos.

Rather than worry about the next ice age, it would be more instructive to consider current changes in ice coverage around the polar regions, as these will certainly influence our lives in the more immediate future. Satellite images and ground observations unequivocally show that over recent years, the Arctic has progressively had far less summer ice than had ever previously recorded by human explorers. This is extremely bad news for polar bears, but possibly a real bonus in that it could open summer shipping routes north of Canada and Russia.

The discussion, opinions, and prejudices on the scale of climate change and global warming are often totally unrelated to evidence, which is well documented, and which can be appreciated by any intelligent person (i.e. without scientific training). Predictions and modelling are far more difficult, as there are many factors and a range of opinions. Even climatologists disagree on the details. Therefore I will not discuss predictive models, but focus just on data that are not in dispute.

In the case of the satellite images of the steadily reducing summer ice cover of the Arctic Ocean, there is no question that they show very dramatically that in this region of the northern hemisphere, warming is taking place. No scientific training is required—just direct observation of the changing yearly photographs. It is clear evidence, even if arguments exist over the cause.

At the opposite end of the world, there are immense glaciers on the Antarctic land mass and also many glaciers stretching out over frozen ocean. Interestingly, because the underlying ocean is heating, it is undermining their stability, so the glacier ice sheets start to flex with tidal movements. One consequence is that large chunks of ice and glacier have broken free and headed out to sea. The only difference between contemporary and former loss of glaciers is the scale, speed, and size of the pieces that are coming from various ice shelves (particularly in the West Antarctic). Satellite data show the current loss is more than 150 billion tons of ice per year. It sounds like a lot of ice, but the earth has a lot of oceans, so in terms of European sea level, this gives a tiny rise of less than a millimetre per year.

Nevertheless, the speed of melting is increasing, so over the next century the change in sea level will certainly influence many low-lying areas across the world. This includes islands, cities such as Venice, and much of Holland. London already needs a tidal barrier to prevent flooding and loss of the underground transport system. It is conceivable, and with current trends probable, that the present Antarctic pattern will result in the loss of all the Amundsen Sea ice, which would raise global sea level by more than a metre within the next two centuries. Other larger glacial areas might follow. In terms of the Antarctic, the confusing factor for predictions is that ice on the solid Antarctic mountain continent is actually increasing in some areas—a feature seized upon by those who do not wish to recognize there is global warming.

A difficulty for the ice scientists, which is not explained in the reports of the popular media, is that there is a need to measure the thickness of the ice, not just the surface level. The entire Antarctic continent is depressed by the weight of the glaciers. Therefore when the glaciers move or melt, the underlying rock can slowly spring upwards, giving a false appearance that the ice surface level has increased. In altitude the surface may have risen, but in terms of thickness it may be less; more complex and detailed data are needed to separate these alternatives.

I suspect this is not the only example of scientific information that has been oversimplified by the media, and indeed may not even be known by those with a scientific background.

Attempts at climate prediction

Climate predictions are difficult, and it is worth remembering that a century ago we had little idea of the driving forces and their consequences. The saying 'Red sky at night, shepherd's delight; red sky in the morning, shepherd's warning' was true, but at the limit of our predictive power for weather. We now have more data, better understanding, and highly sophisticated computer modelling that offers a fairly accurate forecast for maybe a week ahead. Stretching this modelling to really long-term changes is still challenging, even though broad patterns of climate warning are feasible and predictable (only the details are tentative). The very encouraging aspects are that as recently as 1987, when computer power was vastly less than today, the meteorology failed to successfully warn that a small hurricane would strike the UK in October. By contrast, in 2013 the UK weather gurus could not only predict that a major storm was developing in the Atlantic, but do so several days in advance of the event. Further, the modelling indicated correctly where, when, and how powerfully it would make landfall on the American coast. Supercomputers and refined modelling may continue to improve forecasts, but this is an expensive business. In 2014, a planned improvement of just a factor of 13 in computing power was budgeted to cost the UK £97 million.

To most of us, changes in temperature of a few degrees do not seem very profound, as during the course of the year the average daily, and monthly, temperatures in, say, the UK are quite variable, both locally and across the country. Over the last half-century, values have averaged around a range 4–17°C (39–63°F) from January to July, with some years being a few degrees away from the longer-term average. This scale of variation is quite typical in many countries and, for an area as large as, say, the United States, different parts of the nation will have extremely different seasonal patterns and values.

So it is not surprising that there is a difficulty in recognizing long-term changes in the average values. In part, there is a reluctance to consider changes if there are financial implications (i.e. subconsciously reject the possibility of global warming). Slow changes are also difficult to recognize because within our lifespan they may be less than 1°C. In countries where the natural swing in temperature from winter to summer is much more, then the change in patterns are difficult to visualize and appreciate as a long-term trend. Perceived temperature is actually a poor indicator of long-term trends, which are best viewed graphically from detailed records. This requires the skill to understand what is being displayed on a graph.

Among the major reasons that discussions of global warming are so acrimonious is that the general public, and many politicians, have little scientific training (or indeed actively reject anything termed scientific—despite their use of all the modern technologies), so they gain no information (and definitely no insight) when presented with graphical or tabulated data.

A second problem is that for a complex problem, such as predicting the speed and magnitude of climatic changes, there are many variables, a range of views, and distortions from the egotistical scientists who will happily discredit their rivals. For all of us (with or without scientific training), the easy option is complacency—ignoring the entire problem. Alternatively, in most complex discussions we focus on results and views that support what we would prefer to hear. Further, the more affluent are cushioned from climatic changes both at work and in their private lives. They therefore are more likely to only focus on matters that might alter their wealth. Historically, in industrial terms, this has invariably meant more power and more pollution.

Instead of trying to guess if our own experience is clearly telling us there is climatic warming, over and above the normal fluctuations, we need to find a change that is more readily recognized by us without reference to detailed data. In fact, we are far more observant of changes in intensities of storm patterns and rainfall than of temperature. These are related to temperature, as seawater evaporation increases with temperature, and higher values give more energy to storms, and more rain.

So consider the consequences of starting from a modest mid-Atlantic surface water temperature of 21°C (70°F), and then ask what happens if it warms just 2°C (3.6°F), which is well within the effects seen in recent years. This very small temperature shift causes more water vapour, higher rainfall, and more energy transported into the atmosphere. Just a 2° shift can generate an increase of about 13 per cent in total rainfall. So an increase in storms and flooding will be far more obvious than the temperature change. If we are unlucky, and the more serious climate-warming scenarios turn out to be correct, in the Atlantic region, where the weather of both the UK and the US eastern seaboard originate, simple physics says that a temperature rise of 5°C (9°F) will generate around 35 per cent more rain! The extra energy will equally drive more storms, higher winds, and yet more hurricanes. On the western side of the Atlantic, the hurricane season would not just have more powerful examples, but it would extend over more months per year.

Predicting changes is complex, as the very same basic temperature variations influence the way the jet streams swing around in the upper atmosphere. Examples in recent years have shown very clearly that when they do, the USA can have heavy snowfalls and the UK highly varied and energetic weather conditions. Intense sudden flooding in parts of the UK have followed rates of daily and total rainfall that had never been recorded in the 270 or so years of UK meteorological records. The flooding of towns by rivers bursting their banks is partly a consequence of such towns developing alongside rivers, or at the junction of major waterways. Investing in umbrellas and boats may be a good long-term strategy.

Evidence of these changing weather patterns is obvious (at least on the scale of a few decades). For example, in parts of Virginia, the temperature at Christmastime 2015 reached 70°F (definitely not the weather for sleigh bells). However, by January 2016, Washington, DC, had several feet of snow and hurricane-force winds, which were claimed to be among the worst ever recorded in the region at that time of year. The fluctuations emphasize how tracking climate change by our individual experience is remarkably difficult—we need a more clinical overview of weather records.

Disease and plagues

To continue with these rather unfortunate problems beyond our control, we should not ignore biological and botanical factors. Of the many historic examples of plagues that have swept across the world, I have already briefly mentioned two of the most familiar ones. These were the Black Death, a bubonic plague, ca. 1347–53, and the 'Spanish flu' pandemic at the end of the First World War. The population by 1918 was much greater than in the fourteenth century, and there was far greater mobility to transport the disease across the world. In terms of flu-related total deaths at that time, mostly of younger people between 20- and 40-years-old, the estimate is around 20–40 million. The death toll from influenza was considerably more than the number who died from the military actions of the war—equivalent, in fact, to the British population of that period.

Estimates of the numbers of victims of Black Death are quite variable, but some historians cite values as high as 50 million people (of all ages), which would have been around half the European population at that time. Land travel was slow: the typical plague front moved at maybe 2 km per day along the major highways, slower in country areas. Ships were faster: they covered perhaps 60 km per day. Hence ports were key sites for ingress of the plague. Some ports, such as Venice, very sensibly refused to allow sailors to disembark for 40 days to check that they were free of the disease. The practice has given us the word 'quarantine' from the 40-day period. Overall, death on this scale had a major impact on all society. At a time when knowledge was verbally transmitted, the plague not only caused loss of historical information, but also a loss of manual skills of craftsmen.

Numerous other plagues and diseases have erupted across the globe, which in sheer numbers may have claimed more victims, but none so far seem as obviously dramatic as the Black Death or Spanish flu. The really serious difference for the spread of modern pandemics is the ability to move a disease between countries at an unprecedented rate by air travel. Effectively, major cities are globally all joined within a one-day journey. The air conditioning and close proximity of the passengers may actually be ideal for infecting many passengers from a single disease carrier. The air traffic and high mobility can therefore spread epidemic diseases at rates faster than they can be identified, especially since incubation periods before symptoms emerge may often be many days.

A particularly nasty epidemic that is transmitted by close contact is the Ebola virus. The incubation time appears to be highly variable, ranging from a few to 21 days. The only good feature is that the disease may not be contagious until symptoms appear. It is also largely avoidable with good hygiene and isolation of those who are infected, and as of now it has been prevalent only in a relatively limited part of the world. This should not offer any complacency: the downside is that examples of previous Ebola epidemics have had mortality rates as high as 80 per cent.

The final problem is that such diseases continue to evolve. Pandemic transmission via air travel is clearly rapid, and future lethal viruses and high death rates are inevitable. Any new variant that happens to be contagious before symptoms appear will be disastrous.

How bleak are our prospects?

The question may be a good banner headline, but in fact it is too naive. At the individual level, we have a lifespan of normally less than a century. This is unlikely to significantly increase in the foreseeable future. Many factors driving life expectancy are outside our individual control; they will be caused by wars, persecutions, and the types of decay, diseases, and natural events that I have outlined. Therefore the real question is to ask is, 'How well is the entire human species likely to cope with these various difficulties?' On the scale of a major meteor impact or a new ice age, inevitably the total human population will plummet. This does not guarantee that humans will become extinct. If there are survivors and we manage to retain some records, then information and knowledge may survive. This will be more important than just information retained by word of mouth during the immediate aftermath of a meteor strike.

On geological timescales, we are a very new species and still evolving, so, just as for our Neanderthal forbears, we (in the current version of humans) may vanish, and are likely to do so as we gently evolve. Our loss as a species will merely be a natural shift as we are replaced by some new derivative species. This is evolution, and if we are lucky, it may be progress.

Civilization as we currently know it will certainly alter. Therefore the important effort should be to try to control how this change takes place. My personal preference would be in a direction that offers more equality of treatment and opportunities, totally independent of sex, race, and religion, plus maintaining the features that are currently viewed as of benefit to global civilization. This is a major shift and it will need to include an appraisal of resources and all other planetary creatures. This is incredibly idealistic, and unfortunately I also recognize that we have advanced technologically precisely because of, not just intelligence, but the human characteristic features of aggression, power seeking, and personal gain. That part of humanity is unlikely to change. There will certainly be better technology, but we need to examine very carefully if it is desirable in the longer term.

By nature I am definitely optimistic, and therefore convinced (by instinct, not evidence) that humanity will survive even major disasters, and we will eventually evolve into a humanoid form that differs from our present model—in other words, the same pattern of evolution as has happened over the last tens of thousands of years. If we can do this, our intelligence and technological advances may be beneficial.

Certainly, if we are to survive then we need to be aware of possible dangers, plan ahead, and be able to cooperate across the world stage for the benefit of all mankind, not just for local political, racial, or religious reasons. Since this is quite a contrary approach from most human activities, which have all too frequently focussed on power, control of other people, and lands and just plain greed, I am worried.

I will take a simple salutary example of how most of the world has actually behaved when we were driven by greed, profit, and power. Prior to the mechanical items that we put under the umbrella term of technology, a highly profitable route to greater productivity was to use slaves. Probably few major nations have been exempt from this exploitation of their fellow creatures. Since I am British, I will cite the historic behaviour for which Britain has been guilty, but equally I could quote identical inhumane examples from virtually all nations. My example has a slight positive tinge, as at least in Britain we eventually made progress away from this cruel behaviour.

My example is based on our attitude to slavery, and in particular the hypocrisy and despicable behaviour of industrialists, politicians, and, not least, the Church of England. Christian teaching clearly states we should show care and love to all fellow humans. Despite this, Britain was instrumental in stealing some 3 million people from Africa and shipping them to America and the Caribbean to work as slaves. This financial gain was wrapped in claims that we would bring them into Christianity. Most of the home-based users of the products and wealth they generated never considered that their lifestyle was based on slavery. Slaves brought immense wealth to Britain; without it, we might never have risen to be the world-dominating nation of the nineteenth century, or indeed have our present status.

My outrage is that the Church of England was such a greedy and willing partner in this slave trade. It is well documented (in the National Archives) that several hundred slaves in the Caribbean were branded on their chests, with a hot iron, with the word 'Society'. This registered that they were the property of the Society for the Propagation of the Gospel. The Society's treatment of the slaves was to flog them (sometimes to death) and use chains to restrict them. How they could reconcile this with Christianity is inconceivable, except they provided sufficiently vast financial profits that conscience could be ignored.

The British involvement with slavery ran over three centuries. Eventually it was repealed and then outlawed in the 1830s (indeed this progress included aggressive efforts to do so from Christian churchmen). Nevertheless, the final insult to humanity is that when slaves were freed, payment was made to their 'owners', and nothing to the slaves. In current terms, the total payment is estimated at some £20 billion pounds—entirely to the slave owners. The legislation of the 1830s was weak—not until 2010 did owning slaves become a criminal offence in the UK.

The hypocrisy of slave ownership was (and is) widespread. At the time of the American Declaration of Independence, there were some very fine written words and sentiments, which are widely quoted. It is less well publicized that of the 57 signatories, 41 were currently owners of slaves; of the other 16, several had previously owned slaves that they had inherited. The financial reason for ongoing slave ownership was definitely obvious as, in numbers adjusted to the current day, individual slaves were claimed to be worth between $20,000 and $200,000. Indeed, of the 16 non-slave owners of the Declaration, some had no slaves as they were too poor.

In far too many parts of the world, progress has been less successful. Many informed estimates say the current world slave population is around 30 million and increasing, as well as being condoned, or the norm, in some major religions. To put this in perspective, the number is about the total population of some European countries. Despite this inhumane behaviour, we describe ourselves as being civilized. This is definitely still a problem even in the most advanced 'civilized' nations. Media reports of 2016 across Europe repeatedly expose examples of illegal immigrants trapped into slavery as manual workers or sex slaves.

If we are unable to change our greed and exploitation to obliterate such a cancer in society, because of the inherent profits, then no matter how vulnerable we are likely to become, because of natural events and dependence on technology, I do not believe we will ever manage to change our behaviour to plan ahead to safeguard future generations.

Weapons of mass destruction

In addition to natural events, there are possibilities of our destruction from human activities. These include the use of weaponry, and acts of terrorism, that fall under the cloak of 'weapons of mass destruction'. To mention them in this chapter may variously be viewed as despondency or realism.

There are known dangers from the great technological progress in weapons of war, from nuclear to biological weapons, and the power and stocks of such material are immense. If used, they could destroy all of us. This is not a new danger, and most rational governments are aware of the hazards and try to retain control of such weaponry. Unfortunately, the real danger is that, once invented and developed, there is no way back. Therefore acts of an irrational leadership, or even an individual terrorist or psychopath, could unleash global devastation. Many people have considered the outcomes of such weapons with very pessimistic predictions. The views are not limited to cranks or scaremongers, but are often carefully balanced and well considered, as in the book from 2002 by Martin Rees entitled *Our Final Century—Will the Human Race Survive the Twenty-First Century?* (Lord Rees is the Astronomer Royal.)

The viewpoints of how such disasters could be driven vary with current events, with small-scale acts using conventional weaponry or suicide bombs. Far greater damage will ensue if terrorists use nuclear weapons. Such bombs can be delivered fairly easily into densely populated areas. In reality, many people have sufficient knowledge to build nuclear weapons. Delivery does not need to be via a sophisticated missile—it could equally be via ship, train, or large freight truck. The immediate effects would not be global, although radioactive and political fallout from such an explosion would be immense.

Far more difficult to predict are the consequences of biological weapons. Most authors so far have taken a fairly narrow view and considered chemical or biological attacks using familiar items such as sarin, smallpox, Ebola, or other disease vectors that we have experienced in the past. This is a weakness, as within the last decade the knowledge and engineering of biological compounds has made considerable strides forward (normally with worthwhile objectives). But the dark side of this knowledge is that it is equally feasible to develop biological weapons for which we have minimal countermeasures—no vaccines, no treatments.

The restraint from using them at present is that a truly successful attack could spread around the entire globe before the symptoms were recognized, but an uncontrolled spread would equally reach the country of the perpetrators. Therefore, at least in terms of government-driven aggression or religious fanaticism, because destruction would fall everywhere, the lack of control might inhibit more rational users. Our problem is if the same biological vector were dispersed by a fanatical individual, or sect, that wished to destroy as many people as possible.

Whilst we may try to be vigilant against such events, the reality is that if they occur, they are in the same category as major natural disasters. Open discussion may even be counterproductive, as it will offer suggestions for the very events we wish to avoid.

The good news

In the following chapters, I will deliberately try to pinpoint ideas and situations where technology has, or could, cause us problems, but which, if we recognize them, may be manageable. So although this chapter may appear to have had an element of hopelessness and despair (especially in our attitude to slavery), all else in the book will be presented with a brush of hope.

4

Good Technologies with Bad Side Effects

Technological changes within our control

I initially used a simple scenario of a commonly occurring natural event that could readily be disastrous for nations that have advanced technology—unless they have the foresight to take pre-emptive action immediately. By contrast, I subsequently considered some of the spectacular natural phenomena that will be devastating for everyone when they next happen. For these latter examples, I see no point in worrying, as their global effects are rare and outside our control. My attitude is totally different for the contents of the following chapters, as now I want to demonstrate things that were potentially in our control, or were predictable, but nevertheless we failed to see the dangers in time. Worse is that in many cases, we realized there were serious side effects, but we actively ignored them because of immediate gains and financial profits.

My difficulty is not to find such examples, but to pick those that are likely to happen again, or are leading us into an irreversible downward direction. There are very many areas where technologies are rushing forward with truly positive features, and in our excitement we fail to consider those areas that may have unexpected or disastrous consequences. With the benefit of hindsight, the dark side of initially excellent ideas and progress may now seem obvious or ludicrous, but we are not critical enough in looking forward for negative factors and into the hidden dangers of new and exciting innovations. This chapter is therefore intended as a stark reminder that in the past we have blundered, and are likely to do so again. For me, the particularly worrying aspect is that the more spectacular advances may be matched by equally spectacularly bad side effects. Since the reality is that technological progress is increasing ever faster, we need to be actively aware that there are pitfalls, and consider what may be unfortunate outcomes from innovations that initially look excellent.

Beauty, style, and fashion

Detailed records exist, both written and evidential, from nearly 10,000 years; throughout that entire time, there has always been a fixation and motivation to change our appearance, to have fashionable styles of dress, and to conform to the ideal image of our local society. Depending on our preferences, this produces an attempt to look more desirable, more warlike, or more macho than we think we really are. The driving force is image, and this overrides concerns that relate to health or long-term survival. When life expectancy was short this was irrelevant, but now we must think further ahead.

The simplest changes have frequently been to use body and face paints that make us look darker, more colourful, or paler. Paints are effective, but we rarely consider the properties of the pigments and whether there are long-term effects. Creams that made a pale skin, to show we were not peasants working in the fields, used lead oxide (a distinctly toxic chemical), and other colourings for red and black could be similarly unfortunate and potentially toxic.

This is not just a historic pattern: it still exists, although a desire for pale skin may have been replaced by a preference for deep suntans. Certainly it is attractive to Western eyes, but in the rush to have it, many people have gone for rapid UV exposure, either from the sun or from UV lamps. Over the short term, the results are successful, but the skin damage occurs and, particularly with young people, there is a strong link to later life skin cancers. In the USA, legislation now bans children and young people from the tanning studios. With an increasingly large number of older people, there is also an increase in the use of many creams, etc. to hide liver spots and wrinkles. These introduce new chemicals into our bodies, often with unknown side effects.

Discussions of possible links between suntans and skin cancers are quite complex and, because they involve the word cancer, emotive and often ill informed. The general public may not realize that there are two types of skin cancer, one of which (melanoma) is extremely serious, whereas the other, more common type is not life-threatening. Rapid tanning, whether via a sun bed or by lying on a Mediterranean beach, can lead to melanoma. By contrast, a steadily accumulated tan is less hazardous, and although skin cancer sites may appear, they are rarely life-threatening. It was assumed that the UV from the sunlight promotes just vitamin D (which we need), but obviously it does more than this. Recent statistics point to surprising evidence that those people with steadily accumulated tans (even if they have some skin cancer) have a greater life expectancy than the non-tanned. Medical knowledge and data are increasing, but because highlights are packaged into headlines and short articles in the press or TV, the details are lost. Simplistic media statements blur the research results, and comments on safety and benefits can be confusing, or lost. Rather worse is that although medical opinions may eventually change, the original simplistic view is entrenched in the public sector.

Other manufactured skin coloration, via tattoos, is incredibly fashionable in some age groups. As ever with mass markets, the industrial standards are variable; in some countries, dyes are used that can have long-term carcinogenic properties. One obvious downside of tattoo technology is that a passionate declaration of 'I love Mary' is fine until she is replaced by someone else, and unfortunately tattoo removal is not easy. The good news in this case is that Mary is a fairly common name, so replacements are possible. There is also the reality that a tattoo on firm young flesh may look OK, but once ageing sets in, then it is unattractive on wrinkles. Tattoos are a statement of a specific generation, but the patterns and styles evolve, so they become a clear indicator of an age group as the designs and colours go out of fashion. The art work is then a permanent label of a former lifestyle, even if we change our types of friends.

The more drastic reshaping of our bodies with piercing or tribal scar patterns, stretched necks or earlobes, etc. have existed for thousands of years, and within a local context may be desirable. Such treatments do not always fit well in a modern world where people travel freely, over long distances, to very different cultures. Some of the earlier distortions of binding feet or heads to change their shape are also fulfilling localized concepts of beauty or social status, but are distinctly negative in maturity; for example, women with bound feet as children have difficulty walking.

Even modern bodybuilding can look impressive (if not taken to excess), but once the training stops, the appearance is liable to be worse than the original. Because the priority is the image, people tend to overlook that the diets, drugs, and medications, plus the physical efforts used to achieve the figure, are often a disaster in terms of health. Many of the body 'enhancing' drugs are illegal, but nevertheless still in use because initially they are effective. The longer-term effects are far less positive. For example, anabolic steroids are strongly linked to mood disorders, kidney damage, and other changes that are contrary to the original image enhancement, as they can include loss of fertility and sexual function, or even breast growth on men.

Similar aspects of damage to the body result from both excessive food intake and starvation patterns of restructuring. Modern surgery capitalizes on our weaknesses, so people will undergo major and numerous 'reconstructions' to achieve their ideal figure or facial shape. Having watched some of the examples on TV programmes, it is far from obvious to an outsider that the changes were worthwhile, or even an improvement. The remodelling industry is of course extremely profitable and an excellent business for surgeons. These are highly personal opinions, and many patients are very content, but my own view is that in most cases the results look so unnatural that they are a waste of time, money, and surgical skill. The failures are frequently obvious in attempts to avoid the appearance of ageing.

Historic fashions have included immense wigs, complete with poor hygiene and lice, and corsets to produce narrower waists or larger bosoms (or flatter ones). Each ideal shape has been in vogue despite the known distortions of the skeleton and internal organs, as well as the tendency to induce fainting, etc. Every one of these 'advances' has been driven by some new aspect of technology, whether in the use of bones or plastics for corset materials, silicon for breast implants, surgery for facial sculpting, or Botox and filler procedures. The desire to reshape will obviously continue with future generations, and only the target image is difficult to predict. There will continue to be profit for the engineers and surgeons who enable the changes, and failure from the people to think ahead and consider the long-term negative factors, either in terms of health or appearance at a later stage in life.

My comments on reshaping and image have an important message: if we are unable to recognize the downside even of technology that intimately changes our own health and appearance, then I am not surprised if we fail to see problems from technologies that are complex and outside our personal daily experience and knowledge.

Progress no matter what

In our race to introduce new ideas, innovation, more diverse consumer goods, electronic games, and all the gimmicks that are around us in this decade, we show great creativity. Sadly, this is matched by a desire for instant profit plus instant communication and gratification. It may just be human nature, but there are many downsides to this approach that we choose to ignore at our peril.

I see these negative factors in three broad categories. The first is a mixture of purely material activities where we are exploiting and destroying our resources across the entire world (both mineral and human). Many of these are irreversible changes where we have driven species of animals or plants into extinction, or in danger of extinction, by eating them or destroying their habitats. We have equally used them for many derived products, such as rhino horn, tiger skin, elephant tusks, tortoiseshell, and whale oil. We continue to cause deforestation that will never recover whilst there are humans; we have used natural resources such as oil and minerals for which there are limited supplies. We have been just as destructive with the lives of fellow humans. There are many examples of invasions, genocide and theft of tribal lands, carried out by 'civilized' people.

The second category is failing to recognize the historical examples of apparently good ideas and inventions, which have later revealed unfortunate, or even disastrous, side effects.

The third category is where the progress is geared to a limited part of society; the separation between the 'haves' and 'have-nots' is increasing due to advancing technology. Social separation is also more apparent now as media and the Internet disseminate facts and images that reveal the disparities between rich and poor. A few generations ago such sensitive knowledge would have been suppressed. Therefore, this aspect of isolation had never been as obvious as it is now. If I take a slightly arbitrary date of the mid-1990s, when computer technologies gained real power and influence on our lives, then within scarcely 20 years (or less than a generation), modern electronic technologies have driven a divide between the electronically competent and those who lack access to it. Such social splits are already very apparent, but unfortunately they will become far worse in the future.

In this chapter, I will focus on the more practical and mechanical consequences that were, or are, undesirable, and which we failed to predict. Hindsight is of course very easy, but slower progress and more wisdom seem preferable. The story of the race between the tortoise and the hare is a useful reminder.

Later on, I will offer more thoughts on the second facet—how a significant part of the population is becoming isolated directly as a result of our technological changes and progress. To a large extent, they will always be the poor or elderly, as innovative technologies invariably stem from the affluent young who fail to see older generations or the unemployed. Nevertheless, we have a population with increasing life expectancy, which is producing a far greater percentage of elderly people. It therefore seems essential to consider them, not least because the present high-flyers will join their ranks in a surprisingly short time. Without caution and forethought, we could easily be destroying the prospects for future civilizations.

Acceptance of new ideas

One of the more surprising human characteristics has been our attitude to new ideas and 'progress'. There are two conflicting responses, so the overall patterns may not always be apparent until we analyse them. The first response is to a totally new proposal: then we will be negative, and say we do not understand or want it. The second response is to a new product or process has been developed: then we will be extremely enthusiastic, accept it, and totally ignore any potential long-term side effects. To emphasize these two styles of response, I will first consider vaccination, and then, in the next section, give examples of new technologies, goods, and toys that have been invented.

Rather than be excited by items intended for the general good, we often try to block their use. A classic example of the unwillingness to accept new ideas is the attempt to cure, or offer immunity to, smallpox. We now routinely use vaccinations to minimize the spread of many diseases, and in general this is widely accepted as being successful. Nevertheless, for one of the most virulent diseases, smallpox, there was great reluctance to accept the technique. Smallpox was a truly serious disease with as many as 60 per cent of the population being infected and with at least a 20 per cent death rate. Survivors were hideously scarred.

Inoculation had been reported from Turkey in 1721 by the wife of a British ambassador to Turkey, but this was dismissed as being too unlikely to be considered (and, because she was a woman, the observation would have been overlooked by the male medical profession). The less obvious route of gaining immunity by contracting a related disease, cowpox, had been recognized by farmers, but the medical profession did not seriously pursue this, partly due to the social gap between them and farmers.

The medical view started to change because of attempts by Edward Jenner to vaccinate with live cowpox. He had success, but the process was still not immediately accepted; there was even a proposal to legislate against any such vaccination attempts. In defence of those opposed to vaccination, it must be remembered that neither they, nor Jenner, understood the biology of the disease, nor did they understand how vaccination functions. We now view vaccination against smallpox as one of the major medical successes of all time, and the disease was eradicated worldwide by around 1979.

Historic examples of unfortunate technology

The Victorian era offers many examples of exciting and fashionable technology that eventually were realized to have unfortunate side effects. Whilst examples can be cited from all periods of history, the nineteenth century was clearly one of great innovation, imagination, and courage in trying out new ideas. It is therefore not surprising that we can look back and recognize that many of their products were seriously flawed in terms of their impact on the users. Starting with examples from Victorian times has the bonus that we are unlikely to be emotionally involved and so are unprejudiced by current arguments, opinion, and modern commercial pressures. This will not be the case when I present more current examples.

Many of these historic innovations were imaginative and basically sound, but the available materials and production methods made them unsafe. From the twenty-first century viewpoint, their ignorance of chemistry, biology, and physics can seem surprising, and the examples of medical practice are often horrific. Accidents rates were especially high with the new devices based on gas and electricity, not because of the principles, but because of ignorance of the installers and users. So baths directly heated by gas (exactly like a saucepan!) were a luxury compared with boiling water and bringing jugs of hot water to the bath. But the possibility of cooking oneself whilst taking a bath was real; in addition, there was potential for overheating the water and the metal bathtub, scalding, and gas explosions. Such accidents were frequently reported in the newspapers.

Electricity in the home was equally hazardous, as it was a novelty that was very poorly understood. The workmen who installed it were therefore untrained and inexperienced in terms of potential hazards. Even the design of switches, insulation, and cabling lagged far behind the marketing. It is true that Victorian switches had attractive brass toggles and covers, with pretty designs, which may appear preferable to modern plastic packaging, but the metal units were potentially lethal (as indicated on many death certificates).

The more affluent Victorians presumably liked the colour green as they enthusiastically used green patterned wallpaper. One hidden danger was that the colour resulted from the inclusion of arsenic compounds, which reacted with moisture in the atmosphere to give off arsenic vapour. As a consequence, many people were taken ill or died from this hidden killer in their fashionable rooms. Manufacturers of course disputed this link to their wallpaper; at least one major wallpaper designer also owned mines that produced arsenic, so inevitably he would have had a blinkered perspective. The mining activities are never mentioned when praising his work and cultural influence.

Green was similarly popular in the designs of tinted glass, but one of the metals used as a colouring agent was uranium. Uranium provides an interesting green/yellow colour and fascinating, as on a dark night, the glass has a pale ethereal glow. At that time, no one considered the possibility of atomic nuclei, and so there was no concept that atoms might decompose and emit energetic radiation. Uranium glass is certainly decorative, but the owners are unwittingly exposed to the uranium radiation source. Times have not changed, and neither has the physics: I know of modern collectors of such glass who have acquired quite large quantities. In one example, a physics colleague was passing her supervisor's office with a Geiger counter and picked up a healthy signal (actually an unhealthy one). He stored his glass collection under a sofa on which he would rest, as he frequently felt fatigued!

Radiation was also a hidden killer for those who painted fluorescent numbers on watch faces. In order to have a fine-pointed brush, the workers would lick it to provide just a small amount of water. The radium this transferred to their tongues frequently induced cancers.

At the end of the nineteenth century, X-ray sources were being developed. The machines gave nice images of the skeleton, and the users would happily demonstrate the fascinating shadow pictures of the bones. There was no sense of danger to the person from the energy of the X-rays, although in hindsight it should have been realized that many users died of cancers. The complacency towards radiation persisted well into the mid-twentieth century. I recall, as a small boy, seeing and using X-ray sets in shoe shops that were designed so one could look at the way the foot fitted in the shoe. Customers were exposed to quite large radiation doses, but the poor assistants were exposed on every occasion, and so cancers were not an uncommon outcome.

For modern youth, it may seem astonishing that in the 1950s the attitude to radiation exposure was very positive: it was seen as a source of energy for the receiver. Mineral spas and bottled water were proudly labelled with their radioactivity content. Higher values led to better sales. We need to remember that at that time chest X-rays might be required for some job applications, and there was a lack of understanding of the hazards from radiation. Anyone who has seen the newsreel films of atomic bomb tests will realize there was a minimal attempt at protection for the people involved. Workers returned to the sites almost immediately afterwards to assess the scale of the damage to buildings and vehicles that had been left in the test zone. Radiation safety standards then started to set lower allowable exposure limits that dropped by around ten times per decade for the next half-century (an amazing difference in terms of the levels that were considered to be safe).

The reaction to the Cold War and atomic bomb tests caused a complete cultural reversal in attitude to any product that appeared to be linked to 'radiation' or 'nuclear'. It went from desirable to anathema. There was no logic or real understanding by the public, so when a very fine advance in medical imaging came from a new technique called, quite correctly, 'nuclear magnetic resonance', both patients and doctors rejected it because of the word 'nuclear'. This was despite the fact that the diagnostic information was good and there were no harmful effects. Good publicity and a renaming of the technique as MRI (magnetic resonance imaging) suddenly made it acceptable.

By contrast, because X-ray images had been around for 60 years, there was little concern about the use of X-ray imaging, despite the inevitable ionization and damage, plus mutations that are produced in our cells. X-ray examinations can unfortunately never be free of these effects; one of their more common outcomes is development of cancers. This complacency still exists and there is a demand for high-grade X-ray imaging, dental X-rays, mammography screening, and CAT scans (computer-aided tomography), which all depend on X-ray irradiation. The better-quality images with more in-depth information (as in the CAT scans) can only be achieved with higher dosage.

The fact that X-ray diagnostics actually induce cancers is still overlooked by many doctors and patients. This is not a trivial side effect, as various medical sources estimate that around 2 per cent of cancers are actually caused by X-ray imaging. To me, 2 per cent seems high, but the number is the current positive view with state-of-the-art, high-sensitivity X-ray detectors, whereas in earlier decades the downside of examinations was far worse. In hindsight, some early historical consequences were tens of per cent of the patients! For others, 2 per cent will now sound like an acceptable risk, and perhaps it is, but to offer a different perspective, it is worth noting that around one million European women have breast cancers each year, and a 2 per cent value means 20,000 could have been triggered by diagnostic X-ray exposure. From this viewpoint, it is totally unacceptable, especially since there are other non-damaging methods that could be used.

In order to minimize the radiation exposure, image quality can be compromised. Radiation exposure is indeed reduced, and there are two consequences. First, poor image quality may mean we overlook small, early-stage cancers. Second, there is a more significant downside that it is easy to interpret the poorer images to give false positives (i.e. to believe there are non-existent tumours). Some authors claim that there are more false positives than actual tumours detected in many screening programmes. This error leads to anxiety, more examinations and, not infrequently, unnecessary surgery, plus considerable cost to the health service.

Returning now to the Victorians: they were justifiably proud of their indoor toilets, but ignorance of the designs of plumbing needed to cope with released gases of methane and hydrogen sulphide meant that such gases could be trapped and built up in concentration. There was then the excitement of quite major explosions if they came in contact with a candle or gaslight. Once again they provided familiar little newspaper stories. Another problem was that toilets and household water used plumbing with lead pipework, which contaminated the water. Modern understanding not only recognizes the medical dangers of lead, we now know there are many earlier precedents. The wealthy ancient Romans had lead plumbing, whereas the peasants did not. The social benefit of good plumbing was definitely a double-edged sword, as the side effects of the lead have been linked to numerous diseases, madness, and infertility. The plumbing technology may even explain some of the excesses and ultimate failure of their civilization 1,600 years ago.

The Victorians also invented new materials, such as the plastic called celluloid, which was cheaper than ivory, and looked fine, but with ageing became unstable and could be explosive or self-ignite! Because celluloid was used in clothing, the hazard potential was considerable. Once again, many newspaper stories record the accidents. The same fate of spontaneous crumbling or ignition befell the early motion picture industry, as the reels of film were on an unstable, celluloid-based compound. This caused fires not only when running in the hot film projectors, with an intense lamp behind the film, but even in storage containers. The hazard was sufficiently serious that many old films were deliberately destroyed and reprocessed to regain the silver content, rather than preserve historic examples of the industry.

Another popular construction material of the nineteenth century was asbestos. It has excellent thermal properties for insulation and building, but it was many years before it was admitted that the dust particles could cause serious lung damage. Usage of asbestos continued for nearly a century after the hazards had become obvious. Cash outweighed care.

The list of hidden horrors is considerable, but my examples underline that progress can have side effects that are unknown when the new technologies are first used, and, as with asbestos, may be deliberately ignored for decades because the product is effective and cheap.

To offer just one modern example, one could consider a modern composite material made from substances such as aluminium cladding on plastic or polyurethane fillings. These have many excellent properties, and are used on the exterior of futuristic design skyscrapers. Nevertheless, they are potentially flammable, and are thought to have been contributing factors in several skyscraper fires that travelled up the outside of buildings. Twenty years from now, will we look back and say they were an unwise choice?

Victorian kitchens

In addition to hazardous devices for boiling water, cooking, heating, and all the other household chores, many nineteenth-century ‘advances’ in food technology had hidden drawbacks. Two frequently cited examples arose from attempts to improve the appearance of bread and milk. There was a fashion to add alum to bread to increase the weight and bulk. This greatly improved the profits compared with using good grain, and helped with the colour. Alums are aluminium-based chemicals that reduce the nutritional value of the bread and cause many bowel problems (which can be fatal, particularly for children). At a time of high infant mortality, the link to contaminated bread was less obvious than it would be today.

With advent of a good rail network, milk was easily transported from the country into cities. The overall delivery time was still slow, however, so not only was it potentially contaminated with bovine tuberculosis, but it was often sour on arrival. To disguise the taste, people added boracic acid (as recommended by Mrs Beeton). Unfortunately, the consequences of boracic acid range from nausea and vomiting to diarrhoea. It did nothing to inhibit TB; modern estimates suggest that at least half a million British Victorian children died from bovine TB.

In terms of food, there is no room for modern complacency, as the products we buy have been treated and 'improved' with many additives. There may indeed be a list of contents and indications of nutritional value or percentage of sugar, etc. on packaged food, but for most of us the names of the additives, preservatives, and taste enhancers are meaningless jargon. I am probably typical in having very little real understanding of which ones are potentially harmful. The same confusion applies for many 'experts' in the food industry, as opinions on the value of, say, fats, sugar, and cholesterol oscillate between good and bad in ways that seem only to depend on the expert who is backing a particular view.

Another source of confusion is that some additives in Europe are listed with E numbers, but the regulations governing their use differ between countries. Even for food experts, this makes it difficult to know how problematic such compounds really are.

Extra information may not help, as our analytical technologies have advanced, and we can now analyse the multitude of bacteria that exist in our mouth or stomach. Our word association says 'bacteria' is bad news. Our instinct is to have an emotive reaction and try to kill all the bacteria. This is incorrect, as bacteria are essential for our well-being. Just like the word 'nuclear' in the original description of MRI scans, we probably need some alternative that does not have the same emotive overtones as bacteria, and try to invent a new word for 'good' bacteria as distinct from 'harmful' versions. This is still a fairly naive proposal, as the same bacteria react differently in different situations.

Hindsight so far

The very positive aspect of the items I selected here are that, whilst there have been hidden problems, our knowledge base is expanding, and we are able to look back with a mixture of amusement and amazement at the innovations of earlier generations. We may even be impressed that they made so much progress without a detailed understanding of the underlying science. Equally, we are now surprised that they did not recognize many of the negative features of their new ideas and products. Before totally congratulating ourselves, it is worth noting that our grandchildren may make precisely the same comments about us.

5

From Trains to Transistors

Industrial revolutions

Whilst discussing Victorian and nineteenth-century technology, we inevitably think in terms of the Industrial Revolution, and associate it with the major factories, masses of workers, and the big impressive innovations, improvements in manufacturing, and mass production. The outputs from the new technologies included railway transport, steam-powered ships, and steel bridges. Smaller-scale items of bicycles, and flying machines followed, as did motor cars, tanks, and more powerful weapons of war. These were obviously visible and tangible, and we had some understanding of how they were made, and if necessary could attempt repairs and improvements. The details of the materials being used were not that critical. Even for steel, most people realized it contained iron and some carbon, but they probably had no idea of the other additives needed for particular types of steel. This was probably also true for most of the nineteenth-century steel makers. Similarly, brass was made from copper and zinc, but the actual composition and manufacture would not be common knowledge.

This pattern—of using products that were not too sensitive to composition, and could be handled and repaired by the general public—dominated our views on technology. We then have subconsciously assumed that all later advances were in some way derivative products. Instead, we need to rethink—to realize that from the latter part of the twentieth century, there were new and completely different types of industrial revolution. Some, such as advances in metallurgy and chemistry, were just highly refined versions of the initial developments, and these generated exotic metals and materials needed for jet engines, improved chemistry for plastics, and the like. By contrast, the materials of semiconductor electronics and optical fibre communication have required a completely different approach to their development, and one that is alien to the way we consciously function.

The difference is that in this electronic, computer, and communication revolution, we need to make incredible efforts to define the purity of the products, and add into them very precise quantities of carefully chosen elements. Approximations are not allowed, and alternative additives (as may be used in steel manufacture) are totally excluded. The perspective we need is to consider purity control down to parts per billion (one error in a thousand million atoms), and then add into this, in precise regions, new chemical elements at levels that may be as small as parts per million. This level of stringency is outside our daily experience. Indeed, the changes in composition would have been impossible to detect prior to the last half-century.

Some modern electronic devices require such precise control with as many as 60 different elements in various locations. In fact, 60 elements were more than had been discovered and identified in the first Industrial Revolution.

Since I am considering the very unusual challenge of looking at the negative side of technological progress, then this should also include the loss of pleasure and enjoyment as a result of technology. The hugely impressive steam trains, mechanized farm equipment, and fairground rides from the nineteenth and early twentieth centuries may have gone into decay, but they are exciting and, given enough effort, they can be repaired and restored. Consequently, there are hundreds of people who have devoted their time to these activities. The results are theme parks, regenerated steam railway lines, and major destinations for family outings and pleasure. If I try to predict the future and wonder if there will ever be a generation that will attempt to make obsolete computers, so people can play pointless primitive computer games, then I strongly suspect that the answer is no. It certainly could not offer the family-outing-type appeal of the restored steam train journeys.

My aim in this chapter is to try to readjust our thinking to realize that tiny quantities of material can have a major effect on performance and survival of large-scale materials and our environment. The evidence is of course present in every usage of modern electronics or optical fibre communication, but mostly we fail to consider how these items function. Once we accept that tiny quantities of material can dominate the performance of larger systems, it sheds new light on not just modern technology, but also many aspects of the materials we use and our sensitivity to the environment around us. This in turn will focus attention on why we should protect our environment.

I do not intend to discuss the science of the electronics, but I will look again at the role of carbon dioxide and climate. To offer familiar biological examples of sensitivity to such tiny traces of material is easy, and we can relate to them because we already recognize their effects. For example, one can be bitten by a mosquito, develop malaria, and die. The actual weight of poisonous material from the bite may be well under one millionth the weight of the person. Even greater sensitivity is shown in catching a cold from germs floating in the air.

Very positively, we respond to pheromones in sexual attraction, which are highly effective, even when their concentration in the air is as low as parts per thousand million. If the pheromones are successful, then during the next phase of their function, reproduction, we could perhaps add some scientific contemplation. Recognizing that the sperm that fertilizes an egg is far, far smaller than one billionth of the final product, and that it contains more information than we currently can package in an entire library, should give us some scale of the very modest progress that we have made in information technology, and the limitations of our achievements. A sense of humility might encourage us to look at the outcomes of overly enthusiastic acceptance of new technological ideas.

Food—small changes and big effects

Food is a high priority; we not only want enough to survive, but also like to enjoy it. Therefore a good chef will add traces of salt, herbs, etc. to add subtlety and flavour. In terms of volume, these essential additives may only be one thousandth or less of the meal, yet they are crucial. Less obvious (at least to the general public) is that for our health we need minute traces of different chemical elements and compounds to stimulate and regulate our body. In some cases, the quantities are mere parts per million of the total food intake. We may think these are incredibly small quantities and be surprised that they matter. In fact, viewed from modern advanced electronics or optics, they are big numbers. I have just mentioned that in many technologies, such as in the manufacture of semiconductors or optical fibres, it is essential to control trace impurity effects down in the range of parts per billion (i.e. one part per thousand million). To measure and understand this level of sensitivity was unimaginable until the latter part of the twentieth century. It is not easy: it is equivalent to identifying a single individual in a population as large as China or India.

Food flavouring, semiconductor technology, control of our biological processes, including growth, have in common that they are all extremely sensitive to tiny traces of a wide variety of compounds. My non-scientific friends are invariably surprised that such tiny quantities of material are important (except in food), but many of the downside effects of technology are due precisely to such sensitivity. Therefore I will introduce it a number of times, in various types of situation, not least as my scientific colleagues are often just as astonished. Particularly difficult examples to forecast result from residues of drugs and agricultural chemicals, as not only do we not understand their possible effects, but they can accumulate in particular plants or animals (including us).

Openness to new technologies can outstrip our caution, as will become apparent with many of my examples. Half a century of being able to monitor tiny quantities of materials is great for front-line scientists, but there is no way this perception of importance can reach those who develop and use new materials, especially since until very recently we had no idea that such small levels of chemicals could have dramatic effects on our health and survival. Even when the dangers are known, the details tend to be hidden in the medical or technical literature, which is unlikely to be read by the general public. It may also be ignored by specialists if it conflicts with their previous ideas. Additionally, the literature is expanding exponentially, so even experts are likely to miss valuable and relevant new information.

Two other factors exist; the first is fairly obvious, but the second is not. The people most directly working on a product, its applications, and side effects are likely to be those employed by the company that makes the product. They will focus on the advances, not the drawbacks, both because that is human nature, but also because the companies will not want to publicize the downside of their goods and chemicals. The contracts of many employees do not allow them to publish any result unless it is agreed by the company. In extreme cases, there may be whistle-blowers who ignore this, but they can face potential prosecution and may become unemployable by related industries (or worse, if the disclosure is of work conducted by a government body). The second, less obvious difficulty is that both scientific journals and the general media are reluctant to print items that say a process or idea is wrong, or that some previous headline example is no longer seen to be valid.

I have experienced this difficulty on more than one occasion, where I recognized there was a very common serious flaw in the way particular types of experiment were conducted and analyzed. In each case, the mistakes had been made by several hundred authors, including me. Individually, everyone I spoke to agreed that I had spotted long-standing errors, but there was great difficulty in getting into print. The journal editors involved were worried that their refereeing process was being criticized, or there were too many papers in their journals that were incorrect. I persevered, and eventually all the items were published and are now well cited.

The dark side of the Industrial Revolution

The last 250 years have seen an unprecedented rise in industry and innovation. Britain has benefited particularly from it as a result of natural resources of coal and minerals, combined with a cultural and political climate that allowed innovation, entrepreneurship, and expansion of the numerous industries. There is therefore very considerable pride in recognizing the leading role of Britain in so many industrial processes, ideas, and products.

However, the adage 'No gain without pain' seems apt. The advances in the potteries, iron and steel production, energy from gas and electricity, and manufactured goods such as textiles have driven a demand for power, primarily, in the early stages, from coal. Hence there was a surge in the mining industry, especially for coal and iron. In parallel with industrial development, there were a mass movement of people from the countryside to tightly packed housing to operate the mills and centralized industries.

The predictable downside of all this technological progress (with wealth generation for those in charge, and worldwide expansion of Britain's influence) was an exploitation of not only the people in Britain, but also in the colonies, together with devastation and pollution of natural resources across the world.

Pollution was obvious: it was captured in pictures and writing; the industrial heartland was symbolically and literally called the Black Country, and poetically 'the dark satanic mills'. Progress was bought at a high price as the soot, fumes, and toxic gases from burning coal polluted the air in cities and countryside, shortening the lives of everyone, not just those who were mining or working directly in large factories. The tangible evidence of blackened and chemically corroded buildings, as well as vast spoil heaps from the mining and the waste products, has still not totally been covered up.

The pollution in the atmosphere was not limited to the mills and the cities, but was measurable in the woodlands. This produced some instructive side effects. Some species of moth exist in different shades from light to dark. With black-polluted trees, the light-coloured ones were visible, and then eaten by birds, whereas the dark ones survived. It is an obvious demonstration of natural selection as the dark moths had dark progeny. With far less industrial activity producing soot, the air has become cleaner, and light-coloured moths now have a chance of survival. Unfortunately, this is bad news for the tasty dark moths.

Whilst industry flourished, so did the very obvious pollution with slag heaps, acidic rivers, and all the readily visible signs of industrial waste. In parallel, the living conditions in overcrowded cities deteriorated. There were many epidemics and a very high infant mortality rate. Far less obvious, both then and now, was the fact that (as with cooking) the entire living world is sensitive to very small traces of background chemicals that influence our ability to breed and survive as healthy individuals. So by this century, the major chimneys and blackened skies may have vanished in this country, but the more subtle after-effects of past and current contaminants have not.

Pollution was blatant: in London—even without mills, potteries, or ironworks—the Thames was effectively an open sewer. The stench was appalling; cholera and other epidemics were commonplace.

A truly major dark side of industrial progress was the exploitation of humans and other creatures. Many of the industries treated the workers as just another expendable resource. Slavery had been the basis of the classical civilizations of Greece and Rome, and it was still highly profitable in many British colonies. Not surprisingly, the attitude to the working conditions of those in the Britain was not greatly different. Only in name were they free citizens and paid a wage; the majority were trapped into finding employment where it existed, which often meant the mills or the mines. Mining has always been a hazardous occupation, and in Victorian times, many thousands of miners died as a result of underground accidents or lung disease. The situation was accepted as the norm both by employers and workers.

Financial profit took priority, and compensation for industrial accidents was minimal. This was a period when health and safety legislation was seen as a hindrance to progress; industrial accidents were therefore frequent. Even when compensation for accidents was made, the sums involved were small. For example, near Brighton there is well for water that was started in 1858. It is claimed to be the deepest hand-dug well in the world; a sign says a worker was killed whilst digging it. His Victorian widow received 12 shillings and 6 pence (or 62.5 modern pence, less than a US dollar) in compensation. This would have been equivalent to about one week's wages.

More familiar examples of hazardous working conditions are cited throughout all the front-line activities that were driving the Industrial Revolution. Not all of them were reported at the time, in part because hazardous jobs included fatalities, and so they were not newsworthy. One such example was the construction of the original Forth Bridge, 1883–90, when some 57 workers died. A century later a commemorative plaque was planned, but there was no documented list of the names of the workmen, since they were not considered to be important. Bridge construction is still dangerous: in the 1960s' rebuild, there were seven fatalities.

The death rates in mining, mill work, construction, and agriculture were extremely high, and even higher in terms of diseases and disabilities such as deafness or loss of limbs. Britain was not unique in its insensitivity to the care of workers; even in the modern world, the situation has not improved greatly in all countries. Changes in attitudes are probably unlikely as this dark side of technological advance is firmly tied to industrial profits.

The positive aspects are that Britain recognized and responded to these challenges of housing, sewage, living conditions, and safety. In the modern UK, there has been a proliferation of health and safety legislation. This is matched with better compensation rates for industrial injuries. Consequently, both workers and products are considerably safer than a century ago. Whether we have reached the correct balance is less clear, as the growth of the safety industry requires that safety officials generate ever more detailed and restrictive inspections, testing, and certification. So the pendulum of legislation carries a momentum beyond the sensible and necessary. It can then result in excessive restrictions. In many cases, these rules are a distinct hindrance or major expense for careful and competent workers.

Excessive legislation offers an unexpected example of the downside of progress. Additionally, there is the prospect of litigation if a product is unsafe, even during development. This inhibition of exploration and innovation was once summarized by a legal expert who pointed out that 'airplanes, air conditioning, antibiotics, automobiles, chlorine, the measles vaccine, open heart surgery, refrigeration, the smallpox vaccine and X-rays' would all have been totally blocked with current legislation.

Understanding pollutants in our own time

Pollutants are far from simple, as we all feel we have experience and ideas that are frequently in conflict with current UK, European, American, or other global standards. Conflicts because of commercial pressures are inevitable, but in many other cases the official expert views focus on limited sections of a problem; very few people have the ability to provide a truly comprehensive overview.

One recent example is the pressure to use petrol (gasoline) versus diesel for transport. A decade ago, some argued that diesel is preferable as it offered more miles per gallon than petrol, whereas the anti-diesel lobby said diesel engines produced excessive pollutants and were noisy. Both views had some validity, but in the last decade, designers have made enormous reductions in both the noise and the emissions of diesel engines, such that modern 'clean' diesels generate better than 60 per cent less gaseous contamination. This is an impressive advance within such a short time.

But just when the situation started to clearly favour diesel, a new philosophy sprang up; namely, that gaseous emissions from petrol are less important than oxides of nitrogen and particulate from diesel. There is no doubt these are a health hazard in crowded cities; studies suggest they may contribute to 25,000 or more deaths annually. Therefore the proposal is to switch to petrol, ignoring the rise in fuel costs and the increase in pollutant gases from petrol.

This is clearly going to be a long saga, and one which the public will find increasingly complex and expensive. Most will opt for the fuel that reduces their running costs, which will strongly depend on their pattern of vehicle usage. Political overreactions—such as banning diesel vehicles from major cities (as suggested in Paris)—would clearly be an economic disaster for many. In the UK, roughly half of new vehicles are diesel, so a ban on their use would be financially traumatic and unrealistic, no matter what European legislation is in force.

This is an ongoing debate, as petrol is not a perfect fuel. For example, it may contain benzene (which is sometimes detectable from the smell at the pumps). Benzene is a known carcinogenic material, but this feature seems to be overlooked. It is a known constituent; indeed, in Italy the word for petrol is Benzina. So a future media focus on benzene, or other trace chemicals, could happen, suggesting that petrol be banned in favour of diesel. Clearly, they each have drawbacks.

The typical pattern is to focus on one aspect of a problem and draw conclusions that may well be sensible from that specific viewpoint, but unsustainable from others. This, however, implies that the input data to make the claims are correct. In the example of traffic-generated particulate pollution, there is no doubt that traffic is a major source of poor air quality. However, there are two factors that need to be considered. The first is to ask if one can make reliable and quantitative measurements of the level of pollution (probably yes); the second is to quantify the consequences (definitely not, as these are inherently intelligent guesses). Here there is a dilemma, as by careful monitoring it should be possible to have reliable data on the pollution level. The second factor—the number of illnesses and deaths caused—is never going to be accurate, as people and their lives are interwoven with their backgrounds, locality, genetics, place of work, and the amount they travel. Rather worse is that it is bound to be poor science, as it is impossible to make a clear comparison with a situation where the same people are living in a pollution-free environment.

The fallback position is to look at statistics and make intelligent guesses as to the percentage of diseases and deaths that *might* be attributed to background pollution. At this point, it is essential to have the assessment made by people who are unbiased, without any particular agenda. However, since they will have been commissioned to make the study, this is extremely unlikely, and so, no matter what they assess, the results will be presented in a different light. First, the numbers will be cited as definitely saying that London city dwellers are four times as likely to die from pollution as those in the countryside (ignoring all other aspects of their lifestyles); and the publicity will say each borough on average has 100 annual deaths attributable to air pollution. I very carefully have used the definitive and emotive word 'attributable', not the study phrase that should have been used: 'might be attributed to'.

Nevertheless, the press, politicians, and all pressure groups will quote the numbers as facts, in order to pursue their agendas or newspaper headlines. From here forward the values are cited as unquestioned known numbers.

For London, the greatest source of particulate diesel pollution comes from taxis (~50 per cent) and buses (~7 per cent), with less than 20 per cent from private cars. So logical action number one is to ban the taxis and buses! This might not be popular, and re-election of any politicians making such a suggestion is unlikely. Instead, the city administrators might offer a strategy to reduce the problem by replacing taxis and buses with electric vehicles. Viewed locally, this may seem sensible, although the power for the electricity needs to be generated somewhere else (i.e. the pollution is relocated). Furthermore, there are major financial and resources costs needed for new vehicles and materials, not least of which is that there are limited sources of battery materials.

In order to appear to be doing something to reduce pollution from vehicles, a more sensible proposal for many cities might be to try to encourage the use of bicycles. From the viewpoint of health and safety, using data from London could be to ban males from cycling. This seems perfectly logical, since 80 per cent of reported cycle accidents involve men, and the majority of serious cycling accidents occur in urban areas. In terms of UK numbers, for 2013 the number of serious reported cycling accidents was nearly 20,000. Many involve head injuries ranging from 70 per cent in London to 80 per cent of rural fatalities (for 2013 data).

There is also a difference in viewpoint between citing documented accident rates (as from cyclists) and estimating additional deaths caused by vehicle pollution. The estimates are laden with statistics, opinion, and prejudice. In the comparisons between reported cycling accidents and potential reductions in the numbers unequivocally linked to pollution, the benefits of cycling, instead of using vehicles, may make less of a headline than is currently presented.

A proper comparison should also include the deaths caused by mining, transport, and production—as well as pollution—from the attempted switch to new transport technologies. However, we never consider such calculations, perhaps because the nickel, cadmium, and lithium mines needed for the batteries, or uranium mines for nuclear fuel, are in far-distant countries. In the narrow, self-centred view, we might be reducing pollution-linked death rates in our cities, but on a fully comprehensive global calculation, we may be causing the deaths of far more. A blinkered view is much easier on our conscience.

The political claims that with electric vehicles we are using fewer oil-based engines in the cities is also a distortion, as such statements fail to calculate that using electricity to charge the batteries of electric vehicles is inefficient because the conversion from the primary energy source is never perfect. The very best generators, whether power stations, photovoltaic, or wind, vary from as low as 20–40 per cent efficient, so overall we are forced to generate nearly *three* times more energy compared with a direct use of the diesel and petrol fuels in the vehicles. There is therefore a major conflict for those who want both clean city air and less global warming.

Finally, as one who used to cycle regularly on less crowded streets, I see arguments in favour of cycling, but have never considered it for the weekly shopping or in bad weather conditions. When I drive in the highly congested streets of my local town and try to understand why there is now only a single lane for cars, next to the empty zones set aside for bicycles, I am not surprised that pollution levels have risen steeply, as journey times have doubled with the new single-lane system, and direct routes have been diverted, with a consequent increase in journey distances (on both counts this has increased the pollution levels). The single-lane systems are even more problematic when there are breakdowns or minor accidents, as they result in complete chaos. I am probably not alone in thinking that the idealistic so-called green policies to reduce access for cars and use the roads for cycle lanes are very misguided and poorly thought through. Traffic problems do not have a simple, narrow solution. Indeed there may be no solution.

I am being critical of blinkered views, but am also guilty, as I have only discussed the pollutant problems using London as an example. In global terms, the pollution levels there are minor. I have visited many large cities where the pollutant taste in the mouth and lungs is evident within minutes, and then realized why some many of the locals wear face masks. Current estimates say that high levels of city pollution are currently killing more than a million people a year worldwide.

Pollutants and climate change

In my search for the more dramatic examples of the dark side of technology, it is essential that I look at the role of technology on changes in the climate, not least as there are heated and ongoing arguments as to the role played by industrial pollution, and what must be done to counteract global warming. As a scientist, with a career dedicated to careful measurement and cautious interpretation, I am well aware that for such multi-parameter situations, most people will fail to understand the range of information that is available, or what is required. Rather worse is the fact that legislation that tries to minimize production of pollutants, such as carbon dioxide, will cost money. Therefore, many industrialists will intuitively react against any such changes and seize on any evidence that discredits suggestions that we humans are causing global warming. The arguments and evidence that are used often require quite a detailed understanding of the science, and the majority of the population and political leaders (and indeed many climatologists) will lack the expertise to make truly rational and sound assessments. This is not a criticism of their intentions, but the reality of dealing with very complex problems.

Let me be clear that I do not assume that professional scientists are any better in this respect, especially in areas outside their own highly specialized expertise.

To avoid any suggestion of bias, I will start with the tangible evidence that global warming is occurring. For me, unequivocal data are provided by the increasing reductions in the coverage of summer ice across the Arctic Ocean. Viewing NASA images of the shrinking amounts of summer ice over the last 20 years is so obvious that it requires no scientific training to see that there are considerable, and progressive, reductions in the ice cover. Further, this is not just a summer effect, as the Arctic sea ice cover, even in November each year, has steadily dropped, on a noticeable scale of 1.5 million square kilometres since 1980 when the satellite images were available (an area about three times the area of France, or twice the size of Texas, but it is only one third of the entire Arctic Ocean). In fact, the surface area of the Arctic Ocean is one twenty-fifth of the all the oceans of the world. Perhaps one reason we downgrade the significance of this Arctic region is the result of the way we print our maps, as we try to represent a three-dimensional globe on a two-dimensional flat page. Projections such as Mercator are useful for the band of countries level with, say, the USA, but are valueless for the Arctic. Other types of projections distort and reduce the apparent size of the polar regions as they try to include them. Because there is just an ocean around the North Pole we tend to ignore the distortions.

Realistically, all printed map projections are wrong; all are compromises for different objectives. The only good perspective is a three-dimensional globe. Similarly, the focus on maps, originally designed by European cartographers, downplays the scale of Africa. In fact, it is a larger continent than one might guess from a standard atlas. To put it in perspective, the area is roughly equal to the *sum* of the areas of China, Europe, India, and the continental USA. Maybe future technology will offer holographic maps that will redress this imbalance. The predictions that, by the end of this century, the African population will exceed that of the rest of the world seem less improbable once we recognize the size of the continent.

Global warming is also confirmed by detailed records of temperature taken across the planet. The precision of temperature data will vary between countries, and certainly one can always find values that are suspect, but at least for the UK there have been methodical records for around 250 years (from 1777). Therefore if we look at the overall graphical trends, we have some confidence in any pattern that is appearing.

UK data is not global, but it seems to be matched from many other regions, and it shows there is a small but clear trend of a slow rise in temperature over the period of our records (which also happens to be the period we associate with the Industrial Revolution). Most of the warmest years have been in the twenty-first century, with the peak, so far, in 2015. Without access to the wealth of temperature records, it is virtually impossible for us, as individuals, to recognize the changes in temperature. In the UK (or USA), there are major climatic differences across the country, and certainly in the UK rapid and unpredictable daily fluctuations that are quite localized. Most of us are also insulated from the changes by central heating, air conditioning, and appropriate clothing. Additionally, those who travel will be confused by regional and international variations in climate and temperature. The exceptions are people who are more localized, and exposed to the climatic conditions on a daily basis. Farmers are one example, so they will have a greater sensitivity to long-term trends. For them the effects are perceptible: planting seasons and bird migrations have shifted within our lifetime, indicating warmer and earlier springs.

As I have already mentioned, the rest of us may only recognize climatic changes via the number and ferocity of extreme features from rainfall and storms. Certainly the frequency and magnitude of storms has increased, together with higher winds, rainfall, and floods. This is consistent with global warming, because much of the moisture (rain) that reaches us comes from the mid-Atlantic Ocean near the Caribbean. Water evaporation at 0°C (ice) is minimal, but the vapour pressure then increases very steeply with temperature.

I am writing in degrees Celsius, because these are units used globally in science. However, I realize that one of the major reasons many industrialists and politicians in the USA reject the possibility of global warming is not just because it interferes with business profitability, but because they only use, and only think, in the Fahrenheit temperature scale. This is also a difficulty for elderly people in the UK. So for all of them the descriptions and scientific data might as well be in a foreign language. Therefore, I will compare changes in water vapour pressure in both °C and °F. For simplicity I have done this relative to the pressure at 10°C (~50°F). The relative water vapour pressure changes with temperature and goes from 1 at 10°C (~50°F), to 1.4 at 15°C (~60°F), 1.9 at 20°C (~70°F), 2.6 at 25°C (~80°F), to 3.5 at 30°C (~90°F). So over a range from spring to summer temperatures in Washington, the vapour pressure has shot up by 250 per cent.

We do not need scientific training to realize that in freezing weather, the air is dry, but more water evaporates when it is heated, so the humidity rises. Boiling a kettle is an extreme example when water converts into steam. So heating seawater with sunlight also produces more moisture. For the UK and much of the USA, the weather systems are driven by the surface sea temperature in the Caribbean. This has risen by 3°C or 4°C (~7°F) over the last 50 years. Recent peak values have been recorded at values as high as 28°C (~82°F). A few degrees may seem unimpressive, but in terms of water vapour, it is an increase of around 28 per cent! This is then matched by an equivalent rise in the amount of rainfall transported from the sea, and equivalent rises in energy delivered to storms and hurricanes. Before feeling complacent, we should recognize that because there is a steep increase in water vapour pressure with temperature, if there were a further similar rise, then it would generate a 60 per cent rise in water vapour pressure compared to our mid-twentieth-century conditions. This may be the sensationalistic view, but it is not just feasible—it is predictable from the known temperature dependence of water vapour pressure.

Another factor in these discussions is that we assume the oceans are vast and deep, so small changes in the energy trapped in the atmosphere from the sun will be unimportant. However, the weather systems of storms originate from relatively limited equatorial zones; more importantly, it is the *surface* temperature of the ocean (say, the top metre or so) that is critical in terms of evaporation. With this more focussed view on the surface, I realized it needs less than about a 1 per cent increase in trapped energy from the sun to produce the temperature rises that have been reported.

Once again I will underline that we have difficulty in recognizing this upward pattern, as to some extent the changes in day temperatures have risen less than those at night, and the intense storms that are generated have cooled some regions whilst heating others. So, for example, in September 2015 the average temperatures (compared with the twentieth-century average) rose in the USA and Canada by 2.1°C and 5°C (3.7°F and 9°F), whereas Spain was 0.8°C (1.4°F) cooler (i.e. the weather patterns had shifted).

The previous comments are simplistic observations or very basic physics, but we can boost the background data with information stored in ice cores from ancient glaciers. They provide records showing a steady and parallel rise in both atmospheric carbon dioxide (CO_2) and temperature. Both seem to be rising at similar rates and, in the case of CO_2, the behaviour is totally differently from anything in the preceding few thousand years.

At this point our intuition is not very good, as we will assume that, in terms of the total atmosphere, we have added very little pollutant gas from burning fossil fuels. However, in this chapter my aim is to indicate that very small quantities of pollutants, or deliberately added materials, can have major effects across a wide range of science and technology; in other words, intuition may be inadequate.

We know that the Industrial Revolution was a source of CO_2, and the increasing rate of fuel consumption is similar to the rises in atmospheric gas and temperature. We therefore need to see if they are linked, and if so, can we separate out any factors where we contribute to this rise, and if possible, limit it. Foremost in terms of discussion are effects linked to the production of greenhouse gases such as methane and CO_2. The greenhouse analogy just means we are building a chemical layer in the upper atmosphere that allows in sunlight and solar heat, but blocks the escape of energy from the surface of the earth, and so more heat flows in than out. Greenhouses are obvious structures and we can see how they work as there is a tangible glass interface between the inside and the air outside. The fact that CO_2 can do the same is far more difficult to grasp, as we are dealing with an invisible gas. Saying that a doubling in concentration, from around 180 parts per million of the atmosphere to levels near 360 parts per million, is enough to have triggered the climate change, is also difficult to grasp. This will not be obvious or meaningful to many people.

We have very long-term historical records of changes in CO_2 concentrations from glaciers, where there are trapped air bubbles that leave an annual record of the composition of the air for that layer of ice. In the far-distant past, millions of years ago, CO_2 levels were much higher, and so was the temperature. Modern vegetation could not have survived (nor could we). Since the start of the Industrial Revolution, the CO_2 concentrations have steadily increased in a pattern that matches the rise in our temperature records. The logical conclusion is that the two events are linked, not least as more CO_2 provides an efficient greenhouse layer in the upper atmosphere.

CO_2 is a major gas released from the burning of coal, and we are adding it in vast quantities from the prehistoric sources we are mining. The rate of coal usage parallels the CO_2 content in the atmosphere. In sheer volume of material, the coal and oil burning can account for the atmospheric increases of CO_2. Currently, worldwide, we burn some 9 billion tons per year. The pattern has changed as to which countries contribute most, but the upward trend of the total is still rising. For example, China was building some 50 coal-fired power stations per year (i.e. one per week) earlier in this decade.

Different pressure groups will of course say that their favoured energy source will resolve such pollution problems, and we should strive for a non-polluting renewable source of energy. It is idealistic, but not feasible in the immediate future. There are also arguments used against particular forms of energy generation, or factors which are overlooked. For example, there is a fear about radioactive contamination from nuclear reactors (very reasonable), but a failure to recognize that, for equivalent amounts of energy generation, coal-fired stations often release as much as 100 times more radioactive material from different elements in the coal as from a nuclear power station! Most goes into the atmosphere, but some remains as ash. This factor of 100 is the reality because we are using fossil fuels that are inherently contaminated with radioactive elements. Further, we need to remember that radioactivity is absolutely essential for us, as it has maintained the core temperature of the earth. Without natural radioactivity, the earth would be a frozen planet.

Arithmetic for sceptics

The link between global warming and CO_2 production from industrial use of coal and oil, as well as wood-burning during massive deforestation, is certainly clear, precisely because the CO_2 exists in the atmosphere. Whilst transparent for visible light, it is a very strong absorber of long wavelength light (i.e. heat radiated from the surface temperature of the earth). Because we cannot see in the long wavelength, infrared region, we lack an intuitive idea of how absorbing the gas is likely to be, so I will offer a very simplistic analogy for light that we can see.

If we start with a piece of pure aluminium oxide crystal a few millimetres thick, it will look like a clear piece of glass. Indeed, because it is a very hard material, it is often used as a very tough 'glass' for watch faces, so it is something we have seen many times. If we now add some contaminants at, say, 200 parts per million, then it can absorb light and become coloured. The colour depends on the particular material we add to it. If we choose chromium, it will transform into a very nice red ruby; titanium will give a blue sapphire; and nickel a yellow sapphire. So in terms of light absorption, the level of parts per million can be quite effective.

The more difficult question is whether our industry has added enough to make a significant change. Climate change and the role of CO_2 is such a hot political and scientific topic that there are many conferences, opinions, and experts using data that most of us have never seen, and computer modelling that may not be infallible. I am not a sceptic, but I have seen many overexcited claims when reputations are at stake. Therefore, I want to consider if I could attribute the increase in CO_2 gas in the atmosphere directly to our industrial output. My mathematics is trivial. We know both the surface area of the earth and the air pressure on the surface; therefore by multiplying the two together we can estimate the total weight of the atmosphere. The number is quite large at 5,600 million million tons (in alternative notation this is written as 56×10^{14} tons). If we burn one ton of carbon into CO_2 gas, we would produce 2.8 tons of it because we added oxygen during the conversion. For one ton of coal, there will be a little less because of ash residues. Currently we are now consuming 9 billion tons of coal per year.

We need to consider how much CO_2 could arise from the use of oil. Current consumption is around 34 billion barrels per year, which is roughly 5 billion tons. The amount turned into gas, rather than other products, will be somewhat less. Deforestation is hard to assess, in part because it also reduces the take-up of CO_2 from the atmosphere. Nevertheless, the arithmetic says each year we are now adding a mass of CO_2 of at least 2 parts per million. Indeed, it will easily explain why, since the start of the Industrial Revolution, we have gone from 180 to around 360 parts per million. The rate of rise and use of fuels offers a similar pattern over time.

I am therefore convinced that we are responsible for introducing the additional CO_2 into the atmosphere. The only surprise from my simple arithmetic is that the rise has not been greater. Therefore, so far we have been lucky that natural processes are removing some of the gas from the atmosphere. This is not totally good news, as it makes the oceans acidic, and hence will perturb, or destroy, corals and other ocean life. The arithmetic confirms that our industrial input of CO_2 into the atmosphere is more than enough to have driven the rising values since the start of the Industrial Revolution.

The estimates of actual temperature increases that will result are a totally different and a far more complicated problem. The only reliable fact is that we have good records (especially in the UK) that say there is an upward drift in temperature. The changes certainly have a pattern that matches the CO_2 increases, and therefore the inevitable conclusion is that, since we are still adding more pollutants, then the rise in temperature will continue.

A pragmatic comment is that in the early history of the world, there have been periods when the CO_2 concentrations were very much higher, as was the global temperature. These were times when our present crops and creatures (including us) could not have survived. So viewed in terms of survival of the planet, the greenhouse effect is just a passing phase. Viewed in terms of humanity and the world as we know it, there is potential disaster and possible annihilation. Maybe the next intelligent species will do better.

Other greenhouse gases

CO_2 is not alone in having infrared absorption and thus acting as a greenhouse material. Methane offers the same set of problems, but here the uncertainties are much greater, as there are immense quantities of methane trapped in permafrost and cold-water compounds, both of which may release gas as warming continues. If this happens and methane effects cut in as well, then the temperature rise will be rapid, spectacular, and far beyond our control.

Why are we reluctant to solve the problem of greenhouse gases?

The technology and science behind the design of a greenhouse is well understood, and has been so for far more than a century. The science is easy. The sun is hot and emits light over a wide spectral region, not just in the visible, but also out to longer wavelengths termed the infrared (which we can sense as heat). The maximum energy reaches us as visible light. This goes through the glass of our greenhouse and warms the interior. However, inside the warm greenhouse the emission is only at much longer, infrared wavelengths, and glass does not transmit this range of wavelengths. So visible energy comes in, is trapped, and heats the inside. Success: we have a hot greenhouse.

For our survival on the planet, we also need to trap heat to have liquid water and a warm enough temperature. For Earth, the greenhouse glass roof is replaced by CO_2 gas in the atmosphere. This similarly allows visible energy in, and blocks some of the radiant heat from leaving the planet. Without CO_2, we would freeze; with too much, an excess of the energy is trapped, and we will overheat. (Physics is a very simple subject.) Adding more CO_2 from power stations, burning fossil fuels, etc. has changed the earlier balancing rate and trapped more heat. The atmosphere warms up, and this extra energy drives all the changes in the climate that we are currently experiencing. CO_2 has been added by us, but this may soon be a minor problem, as the higher temperatures are beginning to melt permafrost at northern latitudes. This releases trapped frozen methane. As mentioned earlier, methane is also an effective greenhouse gas. Once this methane concentration rises, our global temperature will soar.

We understand the problem; it is serious; and it needs to be addressed immediately, either by technologies that we already have, or by applying intelligence to make improvements and new ways to minimize CO_2 production or remove it from the atmosphere. It is a genuinely urgent problem that we have suspected for a long time, and now have the evidence to realize that we are the culprits causing the change. Therefore, why have we not taken any serious actions?

There are lots of words, international gatherings, and fine talk, but minimal action. Experimental scientists claim we need lots more research (i.e. they want more money to fund their activities). Theoretical ones run different long-term computer simulations, and then argue over the predictions they arrive at for 50 years in the future (the details are unimportant, as they all agree we will have overheated). Industrialists try to discredit all the evidence, because to change the way they operate would inevitably reduce profits in the short term and, for highly paid top management, would hit their salaries. Additionally, they may not understand the science and may not care what happens in 20 or 30 years, as they will have died by then. Others, for political reasons, say the whole prediction is fiction (i.e. a mixture of lack of scientific understanding, a rejection of anything termed science and technology, and a selfish focus on the industrial profits of their supporters). Finally, all of us have a natural reluctance to accept information that we do not wish to hear.

I have listed a range of reasons, as many people have written extensively about different aspects of our inaction and perhaps fail to publicly say that it is not solely motivated by profit. However, that is clearly a very major factor. Many books exist that are long, detailed, and well argued, but few of us are likely to read all the way through them. Nevertheless, for those who wish to be factually well informed and see how politics and commercial pressures have inhibited any action, there are many such books and articles, including Naomi Klein's *This Changes Everything: Capitalism versus the Climate.*

We cannot see greenhouse gases, precisely because they are transparent to visible light. So we do not recognize their importance. So as a digression I will offer an analogy. There once was a beautiful house on a steep hillside with a wonderful view. The rich owner was told by his maintenance engineer that the water consumption was steadily increasing by so many cubic feet per month, and the engineer and the plumber assumed this was an indication of a small leak in the plumbing underneath the floor of the very elegant bathroom. Also, because they were using more water, they suspected the leak was getting worse. The owner prized the expensive bathroom. He did not understand the units used by the engineer, nor any other implications. He was unwilling to spend a lot of money to rip up the floor and search for a leak he could not see. The cost of water was small, and so he just increased the money spent on the water bill. Unfortunately, the water leak produced a sinkhole, and one night the entire house collapsed and slid down the hillside, killing all the occupants. Foresight could have saved them, but ignorance, greed, and a desire not to admit any problems were the all-too-human causes for inaction. Any analogy with global warming is entirely intentional.

Ozone—our shield against ultraviolet light

I want to look at a second material where changes in parts per million have a real influence on our lives. In the uppermost 15–30 kilometres of Earth's atmosphere (the troposphere), the density of gas is low, but it contains a semi-stable oxygen compound called ozone. This has three oxygen atoms instead of the normal two (as in the oxygen molecules we breathe). The poor chemical stability is fortunate, as the upper-atmosphere ozone molecule absorbs ultraviolet (UV) light from the sun and uses the energy to fragment. Eventually the pieces reassemble into more ozone, again by exploiting UV energy. As a barrier to energetic UV light, this chemical cycle had previously been highly efficient, removing more than 95 per cent of the UV sunlight. Without this light shield, the UV energy that reaches the earth is harmful to many animal species (e.g. humans and cattle) where it produces cancers, cataracts, and other diseases. It is also damaging for many types of crop and vegetation.

Unfortunately, ozone is destroyed by reactions with free chlorine in the atmosphere. In the past, atmospheric chlorine was not a problem, as it is not naturally occurring in the troposphere; our technologies, however, have inadvertently injected very large quantities as a by-product of commercial chemistry. Chlorine destroys the ozone via a catalytic process whereby a single atom of chlorine may recycle through many identical chemical reactions, and remove 100,000 ozone molecules.

The chlorine products that are the villain in this scenario resulted from some very fine chemical technology in the USA around 1928 by Midgley and Kettering to produce a superior refrigerant for home and commercial refrigerators. These were termed CFCs (chlorofluorocarbons). As a refrigerant, they were excellent, but inevitably refrigerator units became obsolete, and during disposal the chemicals were released into the atmosphere. The UV at high altitudes released chlorine from the CFC vapour, and that in turn destroyed the ozone. In total, some millions of tonnes of CFCs were produced. We now understand the problem and ban CFCs as well as a plethora of related chemical products, which similarly release chlorine. We may have stopped the source, but ozone recovery is a very slow process, as chlorine persists in the upper atmosphere.

The reputation of Midgley is rather unfortunate, as he also produced a very successful lead additive for petrol that stopped engine 'knock'. Only many years later was it recognized to be a health problem because of lead compounds released into the atmosphere from the petrol.

A final comment on ozone is that we need it in the upper atmosphere to block UV from the sun (i.e. it is a good thing!). However, high concentrations of ozone produced from chemical reactions at ground level are often hazardous for our health (i.e. a bad thing). Ground-level ozone gas is sufficiently heavy (50 per cent more than an oxygen molecule) that it is trapped by gravity near the earth's surface, and only slowly diffuses upwards through the atmosphere. Fortunately, this is equally true of normal oxygen that is essential for life. We should value the gravitational hold that Earth has on our atmosphere. By comparison, on the smaller planet of Mars, which has only 11 per cent of the mass of the earth, gravity could not retain the atmosphere that once existed there.

Twenty-first century technology control of trace contaminants

The real difficulty with chemical reactions such as those between chlorine and ozone is that the chlorine is an activator that drives the reaction but is not consumed by it. So it can react over and over again. Chemists call this a catalyst; similar trigger chemicals in biology are termed enzymes. One of the important considerations with catalytic processes is that very small quantities of material can make a large volume of material react. But precisely because they work in this way, they have often been overlooked. Only in the late twentieth century have we been able to detect them if they were hidden at levels below parts per million.

Parts per million may seem very minor, but in fact is well above the level of purity that must be achieved in many industrial processes. High purity costs money: the purer the material, the greater the cost and development time. So the economics mean that the efforts for extreme purity and cleanliness in production only take place for goods that are of very high intrinsic value. The two most familiar of these are the materials used to make semiconductor devices and the glasses for optical fibres. Both require fairly small amounts of material, but the products sell at very high prices per unit of weight. So the effort and production cost to achieve very pure materials is economically worthwhile.

There are actually two challenges. The first is to find ways to measure and quantify the component elements of compounds or impurities within a material. This must be done down towards detection levels of parts per billion. The second, separate difficulty is to find ways to make such pure materials. During the last few decades, our detection skills have improved, and so we can identify many chemicals that are present in minute quantities of parts per billion. There also has been major industrial progress in first purifying the starting materials and then accurately adding in precisely controlled amounts of impurities into these highly pure materials.

Purifying and then adding controlled amounts of contaminants is not a modern idea. A classic Victorian example was the manufacture of irons and steel from a Bessemer furnace. The strength and hardness of steel are critically dependent on the amount of carbon that is included in with the iron. Early steel making melted the natural mixture of ore and carbon, then removed some of the impurities (slag). The steel maker *hoped* the composition was correct and uniform. This was a poor assumption: because the starting material was variable, the quality of the steel was equally variable. Often the consequences were obvious in that railway lines or bridges collapsed. The major advance was to start with the original ore and try to totally eliminate *all* the carbon (emitting it as CO_2 into the atmosphere). Once this was done, it was possible to add a measured amount of charcoal, and predict the compositional mixture. It worked: the iron and steel were controllable in terms of their properties.

Precisely the same approach is needed for semiconductor electronics, where materials such as silicon are deliberately 'contaminated' with impurities, including phosphorus and boron. Without these trace elements, there would be no semiconductor devices. The electronics industry is intelligent enough not to market their skill in emotive terms of impurities or contaminants; instead they call the additives 'dopants'. This emphasizes that they knowingly added them to the silicon. Exactly the same style of progress occurred with the glass used in optical fibres. Immense effort is made to remove impurities that absorb light travelling along the fibres; these have been minimized down to parts per billion. This process results in glass a million times more transparent than window glass. Nevertheless, the fibre makers then have to add many other dopants to control the fibre properties.

Biological sensitivity to chemicals at levels of parts per billion

There are no normal factors in our lives where we can easily understand numbers as low as one in a billion, so I need some everyday examples. One-in-a-billion sensitivity is equivalent to detecting one individual grain of salt in a kilo of salt. It sounds extreme, but in social terms there have been, and are, politicians and religious leaders who have significant influence over more than a billion other people. The major religions are invariably based on the teachings of a single prophet, and the statements from religious or political leaders, from the Pope to the leaders of India, or China, all impact the lives of a billion other humans. So the importance of one in a billion is not that rare, even if we think it is a little extreme.

Effects resulting from parts per billion may sound incredible, but many animals respond to the presence of impurities at these levels. Since we have learnt of many important examples, it suggests that our sensitivity may equally exist in other circumstances. I am sure many effects will continue to emerge with future studies as our measurement technology improves. One positive example is seen in the way pheromones from female moths are detected. The male moth will sense the presence of a female moth when the concentration of her pheromones in the air is as little as one part per billion. By heading up the concentration gradient, it is then able to chase after her; scent detection can be effective from a kilometre away. This is not exceptional: the homing instincts of salmon and other fish, turtles, etc. that return to specific locations to lay their eggs, rely on an equally extreme sensitivity to scents and trace impurities in the water.

A second animal example is dogs. Their sense of smell is incredibly superior to ours: they can detect other dogs, cats, or humans over distances of 100 metres. Because their scent skills are very precise, we use them in tracking (as in manhunts), to detect drugs and explosives (e.g. in airports), and even in medicine—there are many examples where they have been trained to detect cancers, diabetes, and epilepsy from the chemical emissions given out by the patients. Such sensitivity is currently far beyond our technological attempts to achieve these diagnoses or to make sensors with the same discrimination.

Compared to dogs, humans have a far inferior sense of smell, which is further weakened by the technologies that have driven us into cities and high-population clusters. Those who enjoy a truly rural lifestyle can sense far more subtle perfumes and odours than city dwellers. The same is true of hearing and of other animals. The main observation of ornithologists who study bird song is that, although there are some dialect differences between city and country birds of a particular species, city birds have to sing louder—often double the volume compared with country birds. So for all creatures, including humans, the technological benefits of city life have caused a considerable loss in natural skills, including sensitivity to sounds, sights, and smells.

Scientific techniques, benefiting from highly sophisticated analyses of tiny concentrations of background materials, are increasingly feasible and useful in many unexpected ways. For example, to monitor trends in drug usage in northern Italy, measurements have been made of drug concentrations in the Po river. Despite sewage treatments, traces of drugs survive, and so changes in the background level reflect changes in drug usage in the region. Analysis also reveals the level of drugs ingested by non-users merely as a consequence of using tap water.

Potential future difficulties

As medicine, biology, chemistry, agriculture, and other branches of science develop, we are using an ever more diverse set of chemicals and drugs that do not necessarily degrade when they are disposed of. The Italian river Po example is replicated across the world. Drinking water in London, which has been purified many times, still contains traces of drugs and chemicals that may have entered the water system in Oxford. The contaminant levels may now be detectable, but the cost of purification to totally remove all traces is not only unrealistic but almost certainly not even possible. Nevertheless, many chemicals can both accumulate in certain organs of the body, act in a catalytic role, or both. There are so many examples of such events taking place that I will return to this when discussing some of the dark-side effects of agricultural chemicals. However, the real difficulties may emerge from drugs that we currently think are benign as trace items, but which may eventually be seen to have serious side effects. There are now indications of mutagenic chemicals that do not immediately affect the recipients or their children, but effects emerge in later generations (i.e. grandchildren). If these mutated genes produce serious problems, it will not only be difficult to track back to the original cause, but once a genetic change has taken place, it will not be reversible.

How do we take control?

This would be a rather negative note on which to end the chapter, especially as I suspect the number of examples will significantly increase as we tinker with new biologically active products. The opposite side of this view is that as we gain a better understanding of chemicals and drugs, and their interactions with biological material and the environment, we will be in a stronger position to take effective action to minimize the downside of the reactions. It will also mean that we gain a far better control of what drug usages are needed, and so will reduce the quantities of drugs we use.

6

Food, Survival, and Resources

Our caveman conditioning

Humans are no different from any other animal in that without a steady supply of food, we cannot survive, prosper, and multiply. This essential need has shaped our attitudes in everything we have done as we moved from warm African origins to spread out across the world. Hunger has driven our behaviour, and the survivors are those who won the fights, took the food, and fathered the next generation. This drive to survive is now slightly less obvious beneath a veneer of civilization, and what was once an essential characteristic has unfortunately steadily evolved into the traits of greed and avarice. The main change is that we initially needed food to live, whereas for large sections of the advanced world the attitude is more that we live for food. It is certainly pleasurable, but for many it is an addiction that brings poor health.

Technology in food production has matured from bows and arrows for hunting to satnav-driven tractors that, completely unmanned, can plough the ground and harvest crops. We similarly have developed a taste for exotic produce from all over the world, and expect to be able to have it at any time of the year. This probably does not add to our enjoyment, as older generations say there was great pleasure in having seasonal foods and looking forward to their appearance.

Because we have prospered, the world population has grown, and it is still increasing at a very steady rate. Advances in medicine have also contributed to far greater longevity. So the demand for food is increasing, not least as the Third World countries, not unreasonably, aspire to the lifestyle and nourishment of the advanced nations. In countries and times when child mortality rates were high, and more hands were needed for manual agriculture, the population increases were partially self-limiting. Nevertheless, we are approaching the point where food production, distribution, and availability are stretched to cope with existing high population numbers. We currently survive, but to raise global food standards and production volumes is a severe challenge.

As always in such uncontrolled situations, the facts and opinions of how to proceed can seem quite contradictory. The current population of around 7 billion people is expanding and may well double within the next 25 years (i.e. just a single generation). However, the changes are very diverse in different regions. Numbers are falling in countries that are more advanced and where both men and women have better education, but where women are denied education (often for religious reasons), numbers are increasing. This is despite the fact that in most of the world, the fertility rate (i.e. the number of children born per woman) has declined. Comparing numbers from the 1950s to the present day shows a steady fertility decrease varying roughly as follows: Africa (7→5), Oceania (7→2.5), Asia (7→2), Latin America (6→2), North America (3.5→2), and Europe (2.8→1.8). This is not totally positive, as within Africa a number of very populous countries have increased the fertility rate towards 8 births per woman. If we were to maintain the current world population, then the target number is near 2.1 (with a caveat that we do not increase average lifetimes significantly).

Can we produce enough food?

I like this question because at least I believe there is a real possibility that the answer is positive. My logic is based on the fact that we are currently surviving and that large sections of the advanced nations are not just eating far too much, but happily wasting probably half of the food that is produced. Therefore, with better motivation, a fitter and thinner world population, and increased efficiency in avoiding waste, we should be able to sustain a slightly larger world population than at present. In the short term of one generation, world starvation can be avoided. This gives us a breathing space to educate and rethink how to reduce future world populations, not just to the current level, but to actually reduce it. If we can do this, the quality of life in the underdeveloped nations will improve. This is essential, not least as commercial enterprises want to continuously expand and currently can only achieve this if the population is increasing. New markets would help in this respect whilst we try to change our behaviour. It is clearly contrary to past human behaviour of expansionism and wealth generation by exploiting mass markets. But the current approach is unsustainable in terms of world resources, not just food. So curtailment of production levels is needed. This is a radical change in attitude, but for our survival it is needed.

Historically, we have generally ignored the social factors, and instead focussed on further forward progress, enabled by yet more technological innovations. In our enthusiasm, we have looked for short-term gains for ourselves (especially in the more 'advanced' societies), without worrying about the large sections of the world that are living in poverty and conditions that may be worse than for our Stone Age ancestors. The view is blinkered, but widespread.

How much food do we actually need?

Because I am claiming that advanced societies overeat, I need to justify this and guess at how much we actually need and where are wasting resources. Technologies have greatly aided farming and food production as well as making food a global market. In part this is because food is cheaper if we exploit workers and resources in underdeveloped regions.

We demand more food these days, but estimating the scale at which we now need to produce it, compared with former times, will be contentious. Nevertheless, we can make a guess in the case of meat consumption, as relevant data exist. The numbers are influenced by culture, availability, and religion. From the meat-eating countries, I will use the numbers from 2002. These range from averages below 25 kg (~ 55 lb) per person per year in many African countries, to 80 kg (176 lb) in the UK, to 125 kg (275 lb) in the USA. It seems reasonable to assume that the African numbers are probably typical of our early ancestors, and the US figures are possibly an underestimate of the quantities consumed by many sections of the US population. The changes in technology and agriculture have contributed to the more affluent nations eating five times as much meat per person than our tribal ancestors and more protein than is needed for a healthy diet.

The 'need' for more food per person is only part of the problem, since much of the produce grown in the fields is never used. First, it may just be left unharvested if the local prices are too poor for the farmer to collect it. Second, some crops will be rejected because they do not look sufficiently attractive for the supermarket image. Third, supermarket food will have sell-by dates which are short, and so large quantities are thrown away before being sold. Additionally, particularly with vegetables, the supermarkets tend to sell, say, carrots or parsnips, in large bags. This package looks like a bargain. However, for individuals or small families, they are rarely eaten before they are inedible, so 30 per cent wastage is not rare. This is a general problem that the food that has been purchased and taken home, but is not used—or prepared, but not eaten. Data from the USA may represent numbers from one of the more wasteful societies, but current estimates suggest a total of around 40 per cent of farm produce is wasted. The breakdown is cited as follows: 7 per cent not harvested and left to rot; for some crops, up to 50 per cent of items fail the appearance test for supermarket products; of the items purchased, at least 33 per cent are not used. The net effect is that from growth, to the fraction actually eaten, possibly as much as two thirds is lost, but this will depend a little on the particular produce.

These surveys and estimates probably disguise the real scale of the wastage, as supermarkets will actively, or subconsciously, discourage farmers from admitting how much of their crops are rejected because they do not fit the cosmetic idealized image that the supermarkets want to have on their shelves. I have seen TV programmes showing perfectly edible truckloads of crops being rejected on a daily basis, from just one farm, and entire harvests of some crops being ploughed back into the ground because the supermarket did not want them. From the farming viewpoint, it is a disaster, so many farmers are leaving the industry or going bankrupt, as this model of deliberate wastage is uneconomic. I should not really need to add that it is morally wrong, and totally unacceptable, in a time when large sections of the world are desperately undernourished. There is an even sadder side effect that the suicide rate of farmers is high.

The reasons differ, but effectively the cultures of such a wasteful approach to food are definitely not confined to the USA, but certainly it can be very obvious there. I have eaten in self-service restaurants in the USA where the entry is, say, $10, and the signs say 'All you can eat'. Many customers have a healthy first serving, return for a second, and then collect food for a third, which they invariably leave on the plate. The results are a large wastage of food and many customers who are grossly overweight.

Their obesity causes health problems for which treatments use expensive and sophisticated medical technology, doctors, nurses, and an immense back-up drug industry. Even with a great stretch of the imagination, this is not progress, but merely technology-driven excess. If I am looking for a positive feature, then I could interpret obesity in richer nations as evidence that the world can produce more food than is currently needed, and if we can organize adequate distribution, we could alleviate shortages elsewhere (a very idealistic thought at present, but perhaps not for future generations).

In other countries, the reasons for wastage differ. For example, inability to deliver crops in an edible condition to people is well documented. This arose because of political ideology with a concept of collective farms. These were frequently developed to be excessively large and too far from the cities. So they produced crops, but there was considerable wastage because of poor transportation, lack of adequate packaging, and inefficient delivery systems (such as no refrigerated vehicles).

A separate set of negative features is associated with the destruction of our original landscape. This is glaringly evident in the South American or Pacific jungles, where the original landscape has been destroyed in order to raise cattle and foodstuffs for them. Worse is the destruction, with very dubious efficiency, of replacing food with crops to generate biofuels. Additionally, the crops that are grown may be for the foreign luxury market, and not of direct value to the local population.

Technology and obesity

In the example of wastage I singled out the USA, as Americans are among the leaders in the league tables of obesity, but virtually all the advanced nations have similar problems. The causes are many, most of which are squarely based on our use of technology without sufficient thought as to the consequences. Therefore, far from criticizing our dependence on technological advances across agriculture and medicine, I place the blame for the negative side effects entirely on our human characteristics of greed and gratification.

With rapidly expanding populations, one classic error has been to attempt mass production from vast acreage of monoculture products, such as the immense grain fields of the American Midwest. This was possible because of the development of petrol-driven farm machinery that could work over large areas at a high speed. The technology delivered high output, but persistent replanting of a single crop depleted the soil of key nutrients. Consequently, there were serious side effects of soil erosion. The soil that was farmed was light, and in a region of low rainfall it was initially held together by deep-rooted grasses. Farming destroyed this root structure and left bare soil in the winter, so under high wind conditions, it fragmented and caused the immense 'dust-bowl' problems of the 1930s.

Higher productivity was, and is, also achieved by chemical treatments. Yet again this is initially valuable and effective, but repeatedly working and treating the fields for a specific crop is not. An additional downside is that the run-off from the land has high concentrations of the added chemicals (such as phosphates) that are undesirable, as they have disastrous side effects in the drainage systems, and eventually contaminate the sea and marine life.

Further, monoculture over large areas has negative consequences for the survival of wildlife, including bees and other insects that are essential for crop fertilization. Reliance on a single crop type is similarly hazardous if there is a disease that targets it, as then the entire crop can fail.

It is easy to see how to extend this catalogue of potential damage. The problems apply equally to naturally evolved crops as to ones in which humans have made some genetic modifications to eradicate poor responses or to enhance productivity. The pattern is that new agricultural technical advances invariably offer an initial good return, but over a long period the yields fall even lower than the natural level, and there are many other unfortunate side effects.

A notable success story, but with drawbacks, is fish farming. It is possible to produce large quantities of fish in pens, but inevitably there are diseases that thrive in the fish due to their being contained in high densities within the enclosures. The easy solution is to use biological antidotes such as antibiotics. It is a short-term fix to the problem that leads to many longer-term negative consequences. A long-term danger from feeding chemicals and drugs to animals is that these same chemicals (hormones, antibacterial and growth-enhancing compounds) reach us, and many other plants and creatures, either directly from the food, or indirectly from waste products that are used as fertilizers. In many cases—for example, with fish and chicken—it is impossible to selectively dose those that are sick, so the treatment is more extensive and generously given to the entire shoal or flock.

In all situations of drug or insecticide delivery that cannot be targeted at the creatures and plants that need it, the approach is to use excessive amounts of the chemicals. These are randomly dispersed, and reach many other creatures and plants. Natural selection takes over as a result of the antibiotic usage, and drug-resistant strains then emerge as the dominant variant. This is an ongoing and increasing problem. Transfer to humans should be included in this scenario: normally the first signs appear with the farmers and workers in immediate contact with the region, but the second stage, where it reaches the rest of the population, will inevitably follow. Examples are extremely numerous.

Of yet more concern is that already more drugs are used on animals than humans, and in some countries there is no legislation to limit either the scope or the quantities that are dispersed or injected into the animals. The negative side effects are predictable and observed, but the drugs are assumed to be cost-effective in the short term for the farmer. Excluded from all this is the global view of who has to meet the costs of further drug developments and medical bills for those infected as a result of the applications to animals and crops.

Vast amounts of growth hormones are used to speed up growth or productivity. The immediate positive feature is that, even using the expensive additives, the food seems to be cheaper. Whether or not the flavour and texture are as good is often questionable (e.g. 'perfect' supermarket apples are often tasteless and with a powdery texture). They may well be free of bugs, but this is because self-respecting bugs discriminate against poor-quality fruit. Apples from my garden may be imperfect in shape, but they have excellent flavour and texture. Precisely the same is true of locally raised chicken and pork, which can noticeably differ, and be superior to, the supermarket versions.

Unfortunately, taste and texture are no longer the main driving factors for our choice of nourishment. In part this is because the majority of us are city dwellers, and we do not expect to buy directly from the grower. Further, there are many imports and special deals in the shops (20 per cent off for large volume, or three for the price of two), plus imported foods for which we have never experienced how the original should taste. Supermarket presentation is important for sales, so taste and waste are not their first priorities. Many farm products are grown or engineered to match the supermarket (and public) preference, but really surprising for me was to discover that many crops now contain less than half the nutritional value of the equivalent crops available in the UK during the 1940s, when the war and lack of imports forced us to use gardens and allotments to survive. Modern shopping is easy and costs us little effort, but home-grown food from my gardening friends is distinctly better in taste and texture than the items I purchase.

Shelf life is important in marketing, but this is generally aimed at the appearance of the product, not the taste. Short shelf lives for food are frequently not logical, as in the original preparation of, say, Parmesan cheese or Serrano ham, both products are kept for more than a year before they are considered to be fit to eat (i.e. it takes this long to achieve a good flavour). This longevity is not reflected very well in the sell-by dates. Sell-by and best-before dates can be unrealistically short, and so are potentially a cause of waste. I have seen two-year values attached on both honey and wine. Clearly these instructions are inappropriate. Similar short-term values on tinned food are often equally absurd. In one family, I am amused that the daughters who visit their father carefully scan his cupboards and throw out tins as soon as they reach the best-before date. He never complains, but retrieves them after they leave.

Tinned food can be resilient, and tins of beef from the First World War have been opened and claimed to be palatable. I accept that the throw-away dates differ for the type of content, but they tend to err heavily on the side of caution rather than the common sense of the user. It is of course true that the very early attempts at making sealed cans used a lead solder, and the lead contaminated the food. The consequences were dire. It is thought that the Frobisher expedition to find a North-West Passage through the Arctic used such tins, and the lead induced madness, which contributed to their death in the extreme weather conditions.

To preserve appearance, there are other more exotic treatments of the food. One such is irradiation with fast electrons or gamma rays. The high-energy radiation doses increase shelf life appearance, but the radiation doses that are needed to kill the causes of rotting and decay are frequently so high that there is a change of flavour and loss of nutritional value. A further anomaly is that for some products, the radiation treatments are only allowed in some countries, but not in the UK. However, food that was irradiated abroad can be sold in the UK. This oddity can mean goods are shipped abroad, irradiated, and then returned with an enhanced shelf life. The gains seem bizarre, as they may have spent several days ageing in transit and processing.

One example of biologically induced change (i.e. technological advance, from some viewpoints) is seen in milk production. This has almost doubled over the last decade, but was achieved by placing a strain on the animal, which resulted in a shorter productive lifespan. Clearly, this farming approach is bad news for the cows, but less obviously, it may result in milk of a lower quality and flavour. The hormonal or other biological treatments will be transmitted directly through the milk to us, or indirectly via drainage and water purification plants. Yet again, the contaminants affect humans and other animals. So even if there is no direct link to us via the milk, we may still be contaminated via many other routes. Maybe I have changed, but to me it seems impossible to buy milk in supermarkets that has the same taste as 20 years ago.

Linking such contamination to our health and development is never easy, as there are so many factors involved, but a few clear examples have emerged. For example, at a time when growth hormones were used in high quantities in chicken, some southern European men developed breasts that correlated with their national enjoyment of chicken dishes. Similarly, a recent report from the USA noted that in certain sections of the population, young girls show breast development several years younger than previously. A highly speculative comment is that this may be linked to diet and a food product that has been hormonally treated. No direct evidence has yet been reported, but it seems an obvious potential cause of this sudden biological change.

Farming involves highly interactive and interrelated processes, and because our instinct is to focus on a single problem, the solution that we find may be brilliant for solving it, but consequences and additional problems may not appear elsewhere for many years. The commercial benefits make us ignore the longer-term effects, but once entrenched, it is difficult to stop and change the practice. The delay before we recognize unfortunate side effects can be considerable; for example, in farming and medicine. Chemicals and drugs received in childhood have effects that may not be evident until maturity, or even in old age. I will expand on this in a later section.

More examples of our sensitivity to trace contaminants

Further, we often respond to unbelievably small quantities of chemical signals, so correlations between negative (or positive) effects may be difficult to quantify. Because such sensitivity could not be detected until relatively recently, the manufacturers and users may still be ignorant of the possibilities. I apologize for repeating the point that minute quantities of chemicals can be highly efficient in producing changes, but it is an incredibly important factor that seems unknown to most people, not least as 'it seems like science', and many people automatically assume they will not understand, and so do not even listen.

Catalysis, enzymes, diet, and health

During the last century or so, chemists and biologists have realized that many chemical reactions and biological processes operate via the formation of intermediate steps that lower the energy needed to drive the process. In chemistry, the agents that allow this to happen are called catalysts; in biology they are termed enzymes. Catalytic agents are not consumed in the chemistry of the process, but are released at the end of it. So they can start again. Therefore, very small quantities are highly effective. (I have already cited a catalytic example of CFC agents destroying ozone in the upper atmosphere.)

The underlying idea of catalysis needs no understanding of chemistry, as a familiar analogy of these helper chemicals is seen for schoolchildren trying to cross a busy road outside a school. If they just wait for gaps in the traffic they may never be able to cross. So we provide a catalyst, a school crossing warden. The warden walks out with a sign and stops the traffic, allowing the children to freely cross the road at low speed (i.e. a low-energy process). The warden returns to the side and can repeat the process. The overall crossing rate has increased even though the energy needed is low, and the warden can repeat the process hundreds of times.

Catalysts can be even more successful if, rather than taking part in every chemical reaction, they allow a process to start, which can continue and be self-sustaining. Here my warden starts the crossing by stopping traffic, but once started, a long stream of children can continue to cross even if the warden goes away. There are many familiar household demonstrations of such effects. For example, consider ripening of apples or tomatoes: if we put unripe fruit in a drawer with one that is already ripe, the mature fruit emits ethylene, which stimulates ripening of the others—thus they all ripen faster.

The key feature to realize is that if we produce chemical contaminants that play a catalytic role, their effect on our lives, health, and environment may be totally disproportionate to the numbers involved. Rather than only cite negative examples, I emphasize that it is essential to realize that catalysts play a major role in many technologies. Catalysis was part of our industrial history long before anyone realized that such processes existed. Examples range from fermentation of wine, to making vinegar, soap, and leavened bread. Economically they are important; the chemical industries rely on catalytic processes for large-scale production of compounds as diverse as dyes, ammonia, and nitric and sulphuric acids.

Catalysts can be within the material during the chemical reactions. A 100-year-old example was the discovery that nickel particles in vegetable oils could cause hydrogen to bond to the oils. The chemistry may be unknown to most people, but it is now the basis of making margarine; currently the industry produces some 2 million tonnes per year of these hydrogenated materials by this route. Alternatively, catalysts can be attached to a surface and trigger reactions removing passing gases or liquids. This is precisely how platinum particles (and some other metals) in a car exhaust system break down the toxic hot gases of a car engine. Catalytic reactions are also essential at oil refineries for the fragmentation of oil into different compounds, such as diesel, petrol, and aviation fuel.

Biological catalysts, enzymes, control our bodily processes and have a subtle chemistry that usually includes some metallic content. Therefore the intake of these trace elements is essential for all aspects of our health. Probably foremost among these impurities is zinc. Some 200 of the major enzymes include this metal, and there is a long list of medical problems that occur from zinc deficiency. A survey of processes that use zinc enzyme catalysis could occupy a book. Zinc is needed in aspects as diverse as DNA and RNA production, in defence against viruses, fungal infections, and cancer, as well as in growth and reproductive hormones.

There is a particularly large need for zinc intake during pregnancy. An estimated daily requirement is around 15 mg for normal life, but as high as 20–25 mg per day during pregnancy. Vegetarian diets are often particularly weak in offering an adequate zinc intake, although there can be zinc compounds in garlic, dark chocolate, and seeds. To place these quantities in perspective, 15 mg is roughly 30,000 times lighter than a 1-lb steak, or less than one millionth of the total weight of food and drink most people consume each day. A very rough guide is that 15 mg is the weight of just one thousandth of a teaspoonful of salt.

Despite the small quantities of material needed, a large percentage of the population do not manage an adequate zinc intake. Among the many resultant problems, it has been suggested that this deficiency can lead to paranoia and aggression. Indeed, one Nobel Prize winner has suggested that zinc-rich diets should be distributed in areas of the world (such as the Middle East) where there is an inherent shortage of zinc in the diet, but an excess of ongoing conflict. This may be a very perceptive suggestion, and I certainly think it should be tried as an experiment, not least because under normal conditions, a raised zinc intake does not appear to lead to obvious problems. Medical understanding of the role of trace impurities, even those as critical as zinc, have really only been developed over the last 50 years, so the possibility of dietary and rationally controlled improvements in health may appear soon.

A second important trace metal is magnesium, which has a similar chemistry to zinc. Magnesium accumulates to be the fourth most abundant element in the body, as it is found not just in bones and red blood cells, but also in muscles, nerves, and the cardiovascular system. Intake of a minimum amount of magnesium is similarly an essential dietary requirement; however, excessive intake has a number of serious side effects.

Another trace metal often cited is lead. Lead poisoning is thought to have disastrous consequences, ranging from infertility to madness. It has therefore appeared as an unfortunate side effect of economic progress and civilization as we know it (e.g. lead pipes of the Roman Empire or in Victorian Britain). Lead is also an excellent additive in petrol, but the resulting contamination in the exhaust gases is a severe health risk, so lead has been excluded from the petrol in many countries.

Heavy metals, such as lead or mercury, tend to have obvious deleterious side effects on the body and may even cause brain damage. Ingestion of mercury in the form of methyl mercury is well documented. Mercury was used in hat making; sometimes it is claimed the phrase 'mad as a hatter' is linked to the fact that hatters could develop neurological problems from the vapour of the mercury they used. Mercury contamination has not totally gone away, as some types of dental fillings are known to have toxic side effects.

There are many ways in which different minerals can enter our diet, and some of the routes are unexpected. For example, there was a fashion at the end of the twentieth century for unglazed earthenware pottery used for dispensing health-food drinks such as orange juice. Unfortunately, orange juice is quite an effective acid solvent, and if left in unglazed pottery it can leach out heavy metals and other elements from the container.

Catalysis is not just important for local industrial and biological processes. A highly important catalytic example is thought to occur in star formation in astronomy. The simplest of building blocks for molecules and compounds is to join two hydrogen atoms together to form a hydrogen molecule. In space, the density of isolated hydrogen is incredibly small, so the chance that two hydrogen atoms could collide, and also have enough energy to react at the very low background temperature, is negligible. Nevertheless, material does accumulate into stars and hydrogen molecules exist. So how does it do it? The assumption is that when a hydrogen atom hits a dust particle, it is weakly bonded to the surface and sticks there for a long time. When a second hydrogen atom arrives, the surface bonding allows the two to meet and form a pair. The modern-day equivalent is an electronic dating agency.

How can we recognize when there are delayed side effects?

Most of the chemicals we absorb are via our food and in the air we breathe. So understanding the relevance and potential consequences of catalytic effects from traces of impurities is a prerequisite when looking at food products. They are bound to exist and will have entered our bodies from food grown in different soils, or treated with fertilizers and herbicides. They will also be unavoidable in the liquids we drink. I have already mentioned that tap water contains both traces of chemicals used in purification, and drugs and pharmaceuticals expressed in the output from earlier users of the water systems. (Delicately phrased!) Other liquids, such as wine, owe their distinctive regional flavours not just to the soils and grapes, but also to additives, fertilizers, and the types of barrel in which they were fermented. These taste enhancers may be viewed as contaminants or as flavourings that make them so desirable.

Several of these examples of how traces of key chemicals can influence our lives have emerged relatively recently, and as we gain greater precision in measurement in the physical and biological sciences, the list of examples becomes extremely long. The disquieting feature is that many of the effects on people are dramatic, even lethal. Some of the damage has a long incubation time, and removing the harmful chemicals from the environment and usage is either difficult or economically unwelcome. There is therefore a concern that as we progress ever more rapidly to make new products and chemicals, we may unwittingly be unleashing long-term time bombs for humans, animals, and agriculture, as well as modifying the climate.

Perhaps worse is that up until now many of the chemical-derived technologies have used only simple chemicals in processes where we have a moderate understanding of the science. The recent advances in medicine, biology, pharmaceuticals, agriculture, and genetic modifications are taking us into more complex territory where we will find it increasingly difficult to correlate traces of drugs and delayed changes. The new products may seem marvellous, but unexpected side effects may not initially be apparent, or may be irreversible. The producers will take care to look for unwanted side effects, but animal testing is not a direct model for changes that appear in humans or creatures other than the test species.

A further difficulty is caused by the fact that every human (and animal) is unique, and there will be always be species and humans that do not fit the standard pattern. This is exactly the way drug-resistant diseases survive. The inverse is that a chemical, which is harmless for most, may be lethal for a minority. There is no way we will stop technological progress, but we need to become far more aware of and responsive to symptoms of negative outcomes, especially if they require only small traces of the original substances.

How pure is our food?

The subtle effects from tiny traces of secondary products and drugs are difficult to imagine because they need only occur in quantities which we, the general public, assume are too small to be important, despite well-documented scientific data. So I will ask, 'What level of background rubbish are we consuming in our normal lives?' Impurity contamination is easily measurable, and there is legislation as to the levels considered an acceptable norm. Standards are well defined for products such as food. It is interesting to reflect on what is considered acceptable as background rubbish or cleanliness standards. Our normal experience with 'high' purity materials is often very modest. For example, in school chemistry labs, the chemicals may contain a few per cent of other compounds. In most aspects of our lives, we are very tolerant of unwanted impurities, and tend to ignore them. Examination of labels on any packet of food, or bottle of drink, will happily tell us that our 'pure' natural product contains several per cent of additives to control the flavour, extend the storage life, or things which just happen to be included. Many impurities are never even discussed, so we forget them or tacitly accept their existence.

This delusional ignorance may be a defence mechanism, as we do not wish to think about the fact that our daily bread (or rice) is made not just from wheat grain (or rice). Farmers and bakers are well aware that the grain storage silos inevitably include soil, weeds, and mouse droppings (or even dead mice). Rodent contamination is definitely an imperfection we would prefer to avoid, but it is a problem, as worldwide, rodents contaminate, or eat, 20 per cent of our food supplies. Perhaps we hope that cooking sterilizes the products, but it does not remove the impurities.

The rules governing acceptable levels of inevitable contamination vary between countries, but as an example one can consider a popular item such as chocolate. The cacao beans are left to ferment for a time before processing into chocolate, so are the focus of attention for many animals and bugs. This includes creatures that spend part of their life cycle within the cacao beans. Bug removal would be feasible with very large doses of insecticides, but the chocolate would be unpalatable and dangerous for humans. Accepting some contamination is therefore the only solution. Guidelines for 'purity' of chocolate range from a maximum of 50 to 75 insect fragments per 100 g of chocolate. Other items such as hairs from rodents (e.g. mice and rats) should similarly be kept below about four hairs per 100 g. Just ponder on this with your next piece of chocolate, or any other food!

In general, we have a natural reluctance to consider such inclusions, so have glossy advertising portraying 'pure' healthy products. In our mental rejection of these impurities, we can also forget the bonus side in which the 'natural additives' provide us with a diet containing important trace elements. The focus on 'perfection' in advertising is equally a distortion of reality, which makes us aspire to something that may be quite unattainable in everything from consumer goods to the shape of our bodies.

Before we become too depressed, it is important to know there is a positive side to all these traces of dirt and contaminants. If we are exposed to them from an early age, we are far more likely to develop antibodies to diseases we will be exposed to later in life. There is also evidence that the exposure to soil and contaminants of the real world reduces the chance that we will develop allergies as we grow up. We were designed as a wild animal, and trying to isolate ourselves in a super clean, protected environment is counterproductive. The modern medical profession uses this conditioning response in all of their immunization programmes.

Conclusion

We are busily destroying forests, depleting soils, and generally ruining large swathes of the planet to produce food of secondary or dubious quality, some of which may have long-term side effects. I hope that you will remember this, and still manage to enjoy your next meal or bar of chocolate. In general, I would like to stimulate better attitudes to encourage improved taste and more efficient food handling. Attacking obesity will additionally offer the benefits of a healthier life. I very much enjoy food, including chocolate, but I take part in an active sport (fencing), so am lucky to retain a body mass index (BMI) of 24. Hence I am not currently concerned about my weight. Nevertheless, I see obesity in the city around me, and the devastation and havoc it wreaks on many people, not only in terms of mobility, but in all aspects of their enjoyment of life and long-term health, plus the selfish strain they put on their families and taxpayers.

I realize that citing a BMI value is a useful guide relating height and body weight, but it is in European units of metres and kilograms (BMI is mass in kilograms divided by the square of one's height in metres). So in the UK and the USA, the majority of people will have a problem estimating their BMI. Both countries use height in feet and inches, but in the USA, weight is quoted in pounds, whereas the UK uses stones and pounds. The healthy ideal range of BMI is around 20 to 25. Below that is thin, and above is overweight. A BMI value over 30 is invariably obese for normal people.

Professional athletes may need different guidelines. For example, a 6 foot 3 inch American football player may easily weigh 250 lb, which gives a BMI of 31.3. Technically, this is in the range of obese, but I certainly would not be willing to point this out to such an athlete. For the normal public, a useful guide to attain the 20–25 BMI range for people who are 5 feet, 5 feet 6 inches, or 6 feet tall, is 102–28 lb, 123–55 lb, and 147–82 lb (or, in stones and pounds, 7 stones 4 lb to 9 stones 1 lb, 8 stones 10 lb to 11 stones, and 10 stones 7 lb to 13 stones 1 lb, respectively).

7

The 'Silent Spring' Revisited

Food, survival, and technology

Food and water are absolute essentials for life, so it is not surprising that we have always tried to improve our supplies using ever more advanced technology. We have often succeeded, and developments have been apparent for several thousand years with obvious improvements in selecting strains of crops or breeding larger cattle. Ploughs and agricultural processes also steadily moved ahead in design and effectiveness, so that with the onset of the Industrial Revolution, heavy work and power moved from humans, oxen, and horses to machinery powered by coal and oil. This has usually cut the numbers of workers needed on the land, and most recently the role of humans has reduced even more as farm equipment for planting and harvesting can sweep across the fields without a human driver, relying instead on automated control and highly accurate positioning by the input of satellite navigation.

Similarly, within the last hundred years, science and technology have opened up many new opportunities, with chemicals that act as fertilizers, and herbicides that (at least in principle) target specific bugs and crop diseases. The more recent examples include biological-level changes where humans have actively modified the genetic structure of the plants that we are growing. The aims of this genetic engineering include improved crop strains with higher yields and disease resistance; at first sight, many of them are successful. My caveat, 'at first sight', is necessary, as we are moving into modifications where we have very limited knowledge of an extremely complex genetic area. We have only a beginner's understanding of the subject. Therefore some 'successes' may carry with them changes we are unable to predict, that will have unexpected consequences emerging at a later stage.

This tinkering and invention without detailed knowledge is a typical human characteristic, so there is no point in suggesting we should not do it (we will). Nevertheless, experience in other areas, such as medicine, is littered with unexpected consequences, as I will mention in the next chapter. For example, drugs used for specific purposes have often resulted in apparently unrelated side effects that at the time of their invention were not considered. A truly negative example was the formation of birth defects from thalidomide, whereas the unexpected side effects of Viagra (intended for a heart condition) will be seen in a more positive light. In medicine and genetics, the scope for long-range secondary influences, unrelated to the original objective, is considerable. Many will fall within my definition of the dark side of progress.

As ever in our innovations, we are invariably focussed on a specific problem and rarely have the breadth of vision, or knowledge, to step back and consider side effects. This is very obvious in agriculture, as food is the very essence of our survival. We have made many mistakes along the way and continue to be remarkably inept at learning from the blunders that we made in the past. Instead we continue to assume we should attempt to manipulate our environment on an ever grander scale, and that chemical controls of planting will suffice. We also actively try to remove all inputs of diversity, despite the benefits of insects and other creatures in fertilization and stabilization of our crops and fields.

These are not just examples of stupidity in a technocentric twenty-first century, but are typified throughout the growth of farming over several millennia. Precisely because we have made huge steps in the relevant technologies, such blinkered vision of instant profit and easy farming means we all too easily lose our long-term objective, and in doing so the results are likely to be catastrophic for future generations. Economic and industrial patterns have become global, so our lack of concern is no longer limited to some small localized farm, but impinges directly on farming across the entire world, because now we farm, and transport produce, on a truly interdependent, international scale.

This realization of our shortcomings has been made by many people, so this chapter is intended as both a reminder and a stimulus to recognize the types of error that have persistently been made, and to encourage us to sense the best way to redress the problems, without losing the benefits of high yield and international diversity of foodstuffs. Failure to do so will mean starvation, if not for us, then for future generations.

Underlying reasons we will always struggle to consistently produce enough farm products include changes in climate (not only on the scale of weekly variations in weather patterns), shifts in seasonal features such as rainfall, and also the appearance of new pests and diseases that enjoy the same crops we do. These are often interlinked difficulties, as local climate fluctuations influence the territory covered by insects, and the global contacts mean we do not just import crops and other products, but simultaneously bring in the bugs, plants, and diseases of other countries. Mostly we do not see their arrival, but with imported food and plants, those working in the industry say it is not rare to find not just stray crops and seeds, but also small creatures attached to the produce. Newsworthy items are rare, but arachnophobia is common, so the media will report large banana spiders or even black widow spiders with grapes—but mostly such travellers are unnoticed or unreported.

A more bizarre news item told of a religious group in the UK that purchased a large quantity of crustaceans. The group took them out to sea and released them on the grounds that they should not be captured for food. The group was unaware that the creatures had been imported from Canada, and that this particular species is seen as a problem in the local UK waters.

In general, if the new arrivals like our climate, this is particularly unfortunate, as we rarely manage to simultaneously import the other creatures that have maintained a balance and population control in their host countryside. Hence, they can proliferate here, and we are ignorant of how to deal with them. Introducing the other creatures and insects that prey on our imported pests can be effective, as long as they have a highly specific diet, but if they adapt to the new conditions, then we have merely multiplied the range of the problems.

From hunting to farming

I will sketch out food-related scenarios, which are surprisingly standard in that they have caused successful and dominant nations, or societies, to reach a peak, and then head into decline. Invariably the underlying reason for their political collapse was not because they lost military power, but because they could no longer feed themselves adequately. Warfare and military losses were frequently the consequence, not the cause.

In more ancient historical periods, the food or water shortages were a mixture of poor agricultural techniques, adverse weather conditions, or both. Additionally, a typical drain on the agricultural resources for successful nations has been a rapid rise in the total population and development of ever-larger cities. The pattern of moving from living at the source of the food, in a small farm or hamlet, to being dependent on other producers and transport systems, automatically introduces a measure of vulnerability. Long-term survivors were generally nations that had managed to solve storage problems of food and water to carry them through lean periods. Nevertheless, prolonged droughts (or floods) were invariably the final blow.

A very localized and present example is apparent from the role of technology and transport, which both introduce serious potential weaknesses, for example, at the local neighbourhood level. In the UK, the supermarket food chains keep many types of food that need re-stocking every few days. This puts them at risk if there are snowstorms, fog, industrial actions, fuel shortages, or other events that interfere with the smooth flow of their supply chains. Electrical power loss can similarly result in many difficulties as it will destroy frozen and chilled goods, as well as making it difficult to operate the stores. I recently saw a local supermarket dumping the entire content of some freezers, as their equipment could not cope with a minor heat wave, so it is not just power losses that cause such wastage. Prolonged loss of transport and power causes chaos, and there are daily images of the consequences from news of war zones.

In a few paragraphs, I will offer a very simplistic perspective on how we moved from the hunter-gatherer phase into an agriculture-based one. There are many books and articles offering numerous examples and great detail, and for more in-depth discussion, one book I particularly liked was by Evan Fraser and Andrew Rimas, entitled *Empires of Food*.

Early hunter gathering

Survival by hunting, whether by humans or wolves, means the group needs to be compact and mobile to rapidly follow the food source. The population must also be small and kept that way, either by restricted breeding or infanticide. My instinct would be to suppose that small populations do not greatly modify their environment. This is false: very recent evidence has emerged that the reintroduction of a few dozen wolves into Yellowstone National Park has not only changed the balance of animals living there, but also had a major effect on the vegetation, and an increase in the number of trees that managed to grow past the sapling stage before being browsed. This in turn has changed the run-off and water movements and the stability of the rivers. For most of us, this is a totally surprising result, and only a very knowledgeable and enthusiastic ecologist would have considered such major and positive changes arising from a few predators. The effectiveness of the wolves is of course emphasized by the character of the national park, but the implication that a minor change in the balance for survival of some species has wide-ranging impact on the environment is remarkable. To put this in perspective, Yellowstone National Park is roughly the size of Northern Ireland or Corsica. So for most of us, we would assume a few wolf packs in such a large area would have only very local impact.

Enforcing the maximum pack size is fine for wolves, but for humans there is a very long time needed to reach maturity, so there are advantages in staying in a more settled pattern, and that means there are benefits from an element of agriculture. Early versions may have been to clear small areas of land, grow crops for a season or so, and then move on. With simple tools this was literally only scratching the surface, so basically there was little permanent damage to the soils, and once the group moved along, the land, and even forests, could recover. Better technology, in the form of deeper-cutting ploughs, axes, and saws, started to make permanent changes. The ploughs destroyed root systems and unbalanced the soil composition to greater depth, and of course cutting down the larger trees would make permanent changes to the entire local environment, from plants to animals. Looking at the scale of current forest destruction with modern massive machinery, it is immediately apparent that even the pristine jungles, of, say, South America or the Pacific, are not just being cleared, but totally destroyed forever. The species of plants and animals that have gone will never recover, even if humans self-destruct and leave the land to re-establish itself. In the long term, this is inevitable, but the selfish immediate concern is to allow humans to survive for a few more centuries.

In order to be an effective animal, we are very self-centred. Religions and traditions may claim there are seven deadly sins (lust, gluttony, greed, sloth, wrath, envy, and pride) that are, by religious definition, undesirable characteristics. In excess, I agree, but basically they are the very same motivating forces that have moved us from animals to a moderately intelligent species with great inventive capacity. A rapidly expanding population growth, ongoing warfare, extreme nationalism, acceptance of ideas from leaders rather than thinking for ourselves, obesity, and a desire for luxuries and non-essentials, etc. can all be linked to one or more of the list of seven. But without these characteristics, we would have remained a minor, weak species.

Indeed, all these factors are central to how we have developed agriculture, and also why we so often fail to see the damage that we are doing with our 'improved' methods. For management and simplicity, there are obvious benefits of having large fields producing the crops that we want, or grazing the type of cattle that seem most useful. Because of this, we have removed or destroyed animals that originally existed, and replaced them with our crops and livestock. A classic recent historical example is that within a 50-year period in America, we wiped out around 50 million bison. This conscious destructive act was so that we could grow crops or have ranches for cattle that we introduced there. At the same time, we destroyed or marginalized the indigenous human population and their food source. This may be classic Seven Deadly Sins' behaviour, especially since we overlooked that the overall health of the land has evolved over a far longer period that have humans. Not only are there many creatures that will eat our crops (hence by definition we view them as pests—including the bison), but also they are a food source for other creatures.

Most obviously in the case of the small creatures that eat grain, we cause new problems, as once we destroy the established balance, we discover that some other new pest takes over because there are no longer any natural predators to limit their population. Other factors that we tend to ignore are where we have indirect, but essential, benefits from the so-called pests. Attempts to reduce insect populations are chemically feasible, but in so doing we lose bees and other insects that are crucial for fertilization and pollination. My description is very simplistic, as in reality we also need to consider all the bacteria in the soil, the worms, etc. and their far more complex stabilizing activities in maintaining fertile and continuously productive land that we can farm. Once we destroy this balance, and also extract key minerals that are catalyzing growth, then the ground will become barren and less productive. Eventually it will crumble and turn into a dust bowl or arid desert. A further dilemma is that for crop growth, it is essential to have a steady supply of water, and ways to retain moisture in the soils, as without water, even the best of soils will produce nothing.

Early, very small-scale, local farms, which were self-sufficient, avoided many of these difficulties. Animals were reared, ground was watered, food was eaten, and the waste products finished up as fertilizer that went back to the soil. In many cases, the farmers realized that they could not continuously take a single crop from the same area, so there were attempts at crop rotation or leaving the ground fallow for a season or so to let it recover. The farms were still liable to fail if rainfall changed, so there was generally a move to add irrigation from other water sources. Less obvious is that when we add water for irrigation, it evaporates and leaves residues such as salt. In particular, salt steadily undermines productivity, as it interferes with the chemistry that fixes nitrogen from the atmosphere into the useful biochemistry of soil. Many such historical examples exist—even the Fertile Crescent of Mesopotamia suffered in this way. The changes were slow, and productivity halved over a millennium, so the pattern was probably not obvious within a single generation. Salt problems were of course understood, and used deliberately by the Romans to contaminate the fields of their enemies in Carthage on the North African coast.

Climatic fluctuations, whether long-term or just for a few years, have always had a major impact on survival, and these are not within our control. The Inca civilizations on the west coast of South America oscillated between weather conditions of rain and plenty, and drought and nothing, as the El Niño current in the Pacific came and went. The Incas survived by having exceedingly well-designed storage systems that could distribute food across their nation. Indeed, it was this organization that bound them together, rather than military power. Similarly, some 5,000 years ago, Northern Africa lost its annual monsoon rains and went from being fertile savannah to a desert. The climate shift in this case has sometimes been linked to a minor tilt change of the rotation axis of the earth. A small and inevitable process such as this is sufficient to be locally catastrophic. The effect of the African climate change was visible in human terms by the growth of the Egyptian Empire. Their success and long-term stability is attributable to well-organized food storage to carry them through periods of poor crop yields, plus annual floods from the Nile to fertilize the soil. The biblical writings claim Egypt had periods of seven good years and seven poor ones, so their organizational skills must have been remarkably good.

Not all ancient civilizations were equally successful: the Aztecs were effectively destroyed by a prolonged drought even though they had storage tanks that would take them through their normal half-yearly dry seasons. But these were inadequate for a much longer drought. Many other examples of collapse can be cited from across the globe where changing climate patterns, droughts, or floods hit major civilizations from the Americas to Cambodia or China.

These are dramatic events for the civilizations that survive or collapse, and the changes are driven by natural climatic fluctuations that have shifted the regions that are habitable and fertile from one zone to another. It is essential to realize that our current input to climate changes from atmospheric pollution and the consequent upward drift in temperatures is not of the same character. We are not merely shifting areas that are agriculturally preferable, but instead are introducing a global temperature rise, and this can have a whole new gamut of changes that differ from the natural events. Far back in the history of the planet, the average temperature was very much higher than now, but it could not have sustained the flora and fauna (including us) that we now see and use for survival.

Growth of cities and long-range food transport—early Rome

Farms need labour, but when successful, there is both a need and an ability to support larger populations. Consequently, small hamlets develop into larger towns, and this in turn means there has to be trade and transport of all types of goods. Most importantly, the towns can only survive if food is produced and delivered to them whilst it is still tolerably fresh. The pattern has meant that big towns needed farms within a very large and accessible catchment area. This requires technology both for transport and storage, plus (and certainly not least) many low-paid workers on the farms. Big successful empires focussed on those in the towns and cities, as this is where there was wealth and political power. They generally solved the labour problems with slavery. Democracy may have been a fashionable concept, but it only applied to a fraction of the population (e.g. as in ancient Greece).

Rome is a classic example of a city that grew at the expense of an ever-larger local agriculture, supplemented by imports from colonies. So there was a growth of land and sea transport, which offered the option of more exotic foods than came from the local region around the city. In terms of labour, there are various estimates that to support the ideas that for each 'noble Roman citizen', there had to be 15–30 slaves working in the fields, transport, or within the city. Life at the top was fine, but for the majority it was dreadful.

To control such numbers of slaves, and to exploit the colonies, it was necessary to have a substantial and well-trained army. By default, this was consuming food rather than producing it. Nevertheless, the military were skilled engineers, and as a consequence, the cities in the Roman Empire gained effective water supply systems with impressive aqueducts that have survived to the present day.

The Roman Empire grew over the entire Mediterranean and became dependent upon it. In part this was because their local soil productivity declined steeply with continuous farming. Rome therefore became ever more dependent on imported food supplies. This was an unstable situation, and when in 383 AD there was a major drought in the Mediterranean, there was a consequent famine in Rome. The first attempt at a solution was to expel foreigners and many others from the city. However, this was totally inadequate, and the army of Alaric, a Visigoth, capitalized on it and moved in to exploit the famine to capture the city. His success was of course claimed as a great military victory, but in reality his victory was due to lack of water, over-exploited resources, excessive transport distances, and a nation that was mostly slaves without any motivation to fight. He won, so the story of the victory was his version.

It is easy to be critical of the Romans, as they are far in the past and we have less bias in our judgements. However, precisely the same pattern of exploitation developed throughout all the other major European nations in later centuries, and the pattern continues to the present day.

Repeated patterns from later European nations

From the fifteenth and later centuries, it is easy to find many examples of other European nations rewriting the same historical pattern. Most of the wealthy successful countries entered into voyages of exploration across the globe. They then used organizational skills, piracy, murder, and enslavement of the natives to provide goods for the homeland. Wealth, such as gold, was stolen, and exotic goods shipped to the European markets. They also turned the distant lands into long-range food suppliers for their home cities. Often the motivation was that the crops were not native to Europe. For example, tea was taken from China and grown in the Indian subcontinent. Rubber plants were grown on new continents. Spices were the sole crop allowed in other colonies. This of course brought considerable wealth to the Europeans, but penury for the colonial slaves. Technically, many were not slaves, but their exploitation was for minimal wages, plus their native crops and lands were destroyed to provide monoculture produce of whatever was fashionable back in Europe. Communications from the distant colonies to Europe were poor, so the working conditions were glossed over, as were the high death rates of the workers. As usual, the history written by the colonists differs greatly from the reality.

For example, according to many school textbooks, Columbus was a great explorer. The fact that he was an incompetent navigator is ignored. In fact, he was more than 10,000 miles short of where he thought he was. His social skills were even worse, as he actively enslaved as many natives as he could carry on his ships, and he murdered hundreds of them by burning them. One consequence of his arrival in the West Indies is revealed by modern estimates suggesting that between 70 and 90 per cent of the islands' populations died because of his activities and the diseases brought by his sailors. He is far from unique in having a glorified image that is totally false.

The refined and elegant lifestyles of the rich European merchants, whether Spanish, Dutch, or English, often existed without any thought of the way their wealth had been obtained, but in most cases a social conscience started to emerge and, at least in times of prosperity, attempts were made to improve the conditions of the distant workers. Historically, the underlying difficulty was that the society was dependent on very long-range shipping undertaken by fragile vessels in difficult weather conditions, and persistently endangered by pirates and ships of other nations. The transport was at the limits of the technology of the ships. The producers were poorly motivated and their soils and lands were not maintained. So in every case, the source of wealth was fragile, and collapsed to a greater or lesser extent.

England was certainly one of the more successful nations by the nineteenth century, with a worldwide Empire. It appears, at least as presented in our school literature, that we were not just benefiting from tea from India and Ceylon (Sri Lanka), or from African minerals, gold, and diamonds, but also offering education and a European lifestyle and religion (even if totally inappropriate). The schoolbooks rarely mention the more current estimates that the tea plantations in Ceylon caused environmental destruction and there were tens of millions deaths linked to our imperial expansion; nor do they mention our examples of genocide, as in Tasmania. In reality, the situation differed little from the imperialism of the Roman Empire.

Just as in the case of Rome, the British Empire was destined to be a short-lived period of peace and glory, because of the logistics of transport, depletion of mineral resources, and the declining productivity of the lands we were exploiting. The desire to have goods made of ivory, tiger skin rugs, and timber that takes hundreds of year to mature are all examples of materials and resources that cannot be replenished at the rate they were (and are) being destroyed. Technology, in the form of improved hunting rifles, or more effective chainsaws, are just typical villains in this depressing saga.

This is an ongoing pattern, and by the present day, the United Nations ecosystem estimates are that over the last three centuries, the global area of forest lands has dropped by at least 40 per cent in 25 countries; in a further 29 countries, entrepreneurs and political forces have made the disastrous decision to destroy some 90 per cent of their forests. The land use has switched to growing crops and cattle (including land that provides food for cattle). The driving forces are the usual suspects of major short-term profits, and an increased food supply that is supporting a rapidly exploding world population. It is a pattern that is unsustainable.

Twentieth-century agricultural technology

By the start of the twentieth century, the pattern of growing crops in larger fields had become routine, as tractors provided the power that previously had been limited to horse and ox. The land measure of an acre was nominally the area that could be ploughed in a day. Mechanization has increased this by a hundred-fold. The obvious side effect of larger fields was the rapid depletion of nutrients from the soils, because without a mixture of crops, there were fewer mechanisms to regenerate the biochemistry of the soils. Technology was flourishing, particularly in chemistry, and there were concerted efforts to replenish the nitrogen that was consumed by crops. One route to this was the production of ammonia to add to fertilizers. A successful chemical process was devised by Fritz Haber and Carl Bosch. It was of immediate value in agriculture and the scientists won the 1920 Nobel Prize for chemistry.

Interestingly, Haber was equally determined to find ways to help the Germans to a rapid victory in the First World War and to save lives (i.e. from all the armies involved). He therefore invented a chemical poison gas that was first deployed at Ypres. After the war, he returned to his main priority of the chemistry of agricultural processes.

The need for more food was clear, and the major companies found it easier to work with very large, single-product farms and, wherever possible, to eradicate anything that might compete with the growth of the plants. This was clearly an ill-considered approach, as it was putting all the eggs in one basket. The difficulties in recognizing the dangers were in part economic, in part social. Focus on a single crop is a dangerous strategy, as seen, for example, in the potato problems in Ireland in the 1840s. They were growing an excellent hybrid at the expense of all other varieties, but were struck by a fungus that particularly liked their favoured type of potatoes. Without alternative types of potato, the food supply collapsed and the famine of 1845 ensued—with consequent political turmoil for the nation that has rumbled on into the present time.

A parallel example was seen around 1940 with the development, by Norman Borlaug, of a superior strain of wheat and maize with stronger stalks to support more grain. His solution was to find a dwarf variety, as this could support the larger heads of grain. It was financially successful, but not perfect: although the head was larger, the nutritional value was less than for the taller species. Nevertheless, since it was highly marketable, it moved into pole position of a monoculture species crop. The downside of this technological progress is that a single strain could be wiped out by a disease that targeted it, whereas a diversity of strains would inevitably have survivors resilient to any given disease.

Technologies in agriculture have generally been introduced with good intentions (as well as for profit); the undesirable consequences emerged from lack of foresight, narrow viewpoints that focussed on a single feature, and concepts that were easy to implement from the viewpoint of marketing and distribution of agricultural products.

Not only chemistry, but also biology and physics were advancing at an unprecedented rate during the twentieth century, and all were stimulated by a vast range of inventions during the Second World War. The destruction from the war put intense pressure on all countries (winners and losers) to provide food, and technology seemed an ideal contributor. Further results of the war included strong pressures of control from governments, a degree of secrecy stimulated by military activities, and societies that would never question decisions from government, industrialists, or those in privileged strata of society. In part, this was ignorance on the part of the farmers and employees, and in part it was a cultural attitude still present in the mid-twentieth century. In the UK, there was an ordered, stratified society in many respects, in which one never queried decisions made by those who were assumed to be more knowledgeable. For example, a decision by a doctor or priest was rarely questioned by the general public. Further, the involvement of complex science inhibited lay people from criticizing new technologies, and the major chemical and agriculture businesses were totally blinkered. Looking back half a century, the situation is scarcely believable for us now, as we are so used to ready access to information, at least in most Western countries.

However, the value of Internet access to information is probably overrated, as in many nations it is blocked for political or religious reasons. Further, the range of Internet and electronic entertainment material is vast, and this can distract us from any serious information retrieval. Also, many people will only look at sources and websites that confirm their existing views (or prejudices). Indeed, many sociologists consider that social contacts sites actually isolate subsets of people, as they each form a sufficiently large community that they are self-supporting in their views (no matter how extreme or ill-founded they are). Just as in the 1950s, there is still a difficulty accessing and understanding information of a technical nature, unless one is trained. There is even more difficulty deciding between conflicting interpretations. I of course doubt that these are problems of anyone who has read this far through this book, but such people are exceptional!

Nevertheless, we have progressed a little and have developed far better and more realistic critical attitudes towards the wisdom offered by 'experts'. We have also moved to a society where there are many people with a modest breadth of scientific knowledge, which is freely communicated and accessible through the media and Internet. This should, and has, allowed us to be more confident in challenging practices that are questionable.

The bombshell of 1962

The unquestioning complacency and acceptance of the scientific progress offered by the chemical and agricultural industries was shattered in 1962 by the highly detailed and well-documented book by Rachel Carson called *Silent Spring* (the reprint of 2012 is still powerful reading!). In it, she highlighted the technological downsides of many of the practices that were then current. She began with applications of insecticides and herbicides that were being used indiscriminately. Her first example was DDT, which indeed is a very effective pesticide—it was initially used to attempt to reduce mosquito populations, and hence malaria. Even today malaria kills around one million people per year, so clearly this is a significant disease. In the enthusiasm to suppress malaria, however, the highly negative side effects of DDT in other types of usage were ignored.

The chemical industries were geared up for mass production as part of the wartime efforts, so inevitably they were seeking new mass-market applications, including agriculture. Consequently, vast quantities of pesticides were being manufactured and applied, not just by specific crop spraying, but by aerial sprays from low-flying planes. This style of technology was rapid, but totally random, as it covered far more area than was intended or needed. The USA alone used in excess of 300,000 tons of pesticides, just in 1962. To Carson, and to most modern critics, the very obvious downside of this crude approach is that the pesticides not only killed pests, but also every other creature that came into contact with the chemicals. Doses were high and poorly controlled. Further, the pesticides often were retained for a considerable time, not only in the plants and the soil, but also in the creatures that ingested them from the spraying or by eating contaminated products.

The real mistake was that such crude chemical attacks caused the death not just of specific pests, but all the insects and other wildlife that had been helping to maintain a natural balance. Once destroyed, the process was irreversible. It was also an ineffective method as, without natural predators, the creatures that fed on the crops (nominally pests) could re-establish themselves very rapidly without predators. This is a fairly obvious blunder, because insect life cycles are far shorter than those of birds or small mammals that would normally have been maintaining a natural equilibrium among various species.

Carson cited numerous examples of data that showed, even where the sprayed concentrations were well below parts per million, that many chemicals accumulate in the bodies of animals: subsequent analyses of corpses included organs with values of many thousand parts per million. This is a thousand-fold concentration enhancement as a result of biological reactions, so this totally invalidates testing at the levels that were being applied. Indeed, this style of problem is just as serious today. The results of contamination were death or sterility for many animals, and extinction of various species. All such data were discredited by the agrichemical companies as pure coincidence. They also organized a campaign to discredit Carson's entire work.

Fortunately (perhaps an odd choice of word), people who were using the sprays, or received high doses, were also taken ill or died. Human deaths then opened a way to gain attention for the general public and commence impartial studies and analyses of the effects of the pesticides and herbicides. At the time, in 1962, there was far less understanding that many chemicals can play a catalytic role in biochemistry, and that, once ingested, they could react to produce damage and diseases that would never have appeared in trials on plants for which they were initially intended.

Trials on the toxicity of pesticides on humans were of course never attempted, although they would have revealed a wide range of effects, from temporary to permanent disability or death. There should have been a greater awareness of the dangers, as even then some other species showed extreme sensitivity; for example, young shrimps were killed by insecticide concentrations below one part per billion. From the shrimp perspective, this is a volume of insecticide of one cubic centimetre (i.e. baby shrimp size) diluted into an Olympic-size swimming pool. Note that current analytical techniques can often do one hundred times better than this, so in the future, far more of these problems will be identified.

Genetic time bombs

Unknown in the 1960s was that many chemicals can interfere with DNA (not least because knowledge of the structure of DNA was only just emerging, and the implications of the structures were still unknown). The possibility that agricultural chemicals could cause genetic changes was unexpected by the public. We now know this does happen, but even so, most people are surprised to learn that the changes sometimes do not manifest themselves for several generations. This is seriously perturbing: although the data showing such effects have been gathered for creatures with a short lifespan (as is standard practice in biology laboratory work), there is absolutely no reason that the same delayed, time bomb–type changes may not have been initiated in longer-lived species, such as humans. Examples of drugs that cause human birth defects are well documented, but events that do not register until they hit grandchildren, or great grandchildren, are extremely worrying. Our concern should exist, not least because the major explosion in biochemistry has taken place over the last 50 years, so such genetic changes will scarcely have had enough time for their effects to be apparent. Detection of long-delay genetic changes will be difficult, and associating them with their true cause, virtually impossible. This is not a reversible path, so future generations will carry the new genetic material.

Modern analyses of genetic coding are now routine practice, and, as I have mentioned, detection of impurities and other defects in chemical structures are now feasible down to the level of parts per billion. As this level of detailed scrutiny becomes routine, one can expect many more disclosures of poisoning from pesticides, herbicides, and medical drugs that were initially applied in good faith.

Whilst DDT was a familiar insecticide, there have been many hundreds of chemicals used in agricultural applications. Some, such as Dieldrin, are at least 40 times as toxic to humans, and are sufficiently rapid that they have been used as a nerve gas. Before being critical of the chemists who develop these compounds, we need to recognize that they are often working very hard to attack a particular agricultural or medical problem. For them, success is a compound that does the task they were focussing on. Even laboratory control and careful technology may not reveal any difficulties, either in the short or the long term. Nevertheless, the real field applications, distribution, and commercial profitability can totally distort their successes and lead to highly undesirable outcomes.

Chemicals that may seem ideal for agricultural usage, where we have tolerable control of the environment, and other chemicals can have results that are far less predictable when they are taken up by other creatures. The literature on unexpected side effects of drugs is extensive. The clear message from the medical examples is that results can be extremely variable. No two humans (or other creatures) are truly identical and have the same responses. Additionally, more than one drug (or herbicide) may have been used, and the combination effects add more complexity and variability. If one reads the literature on many well-established modern drugs, there will often be a long list of side effects, which have affected just a small percentage of the users. Indeed, the list of side effects may be greater for the more widely used prescriptions.

Just as chemists made nerve gases in the First World War, there have been many subsequent products designed for warfare. A particularly well-documented example is Agent Orange, which was sprayed as a defoliant during the Vietnamese war. More than 20 per cent of forests in South Vietnam were sprayed, plus an estimated 10 million acres of arable land. To be effective defoliants, they were used in high concentrations, often hundreds of times greater than deemed safe for human exposure. The toxicology studies are totally damming on its effects on humans, and there was a high incidence of stillbirths plus leukaemia, etc. in children (and later generations) of Vietnam veterans in the USA (together with illnesses of the soldiers). Within Vietnam, the situation was far worse: government and Red Cross estimates initially put the total of disabled Vietnamese people in excess of a million. Loss of fertility was also evident. With subsequent human defects and problems for later generations, the estimates of disability are much greater. This is one of the clearest cases of a very dark side of advanced technology.

How successful is a monoculture with pesticides?

Large tracts of land devoted to a single crop may well be ideal for massive-scale farming and harvesting, but they are still going to be vulnerable to pests and degradation of the soil with repeated usage, so we are back to the problems associated with herbicides, pesticides, and fertilizers. The real question for world food production will not be, 'Are they economic in the short term?' (clearly they are, or the approach would not be used), but 'Are they also a long-term solution?' In parallel with this, we need to ask what are the environmental drawbacks and dangers from new diseases, and what are the risks of genetic abnormalities (either of the crops or the animals and humans who eat them, or merely come in contact with them).

Monoculture approaches may initially offer high-yielding crops, but it is very short-sighted to assume that we can discard all the other varieties that have evolved over many millennia. Just as with computer documentation, there is a danger that rushing into new technology will lose many valuable alternatives. For agriculture, if climate or disease patterns change, the apparently less effective versions may be preferable to the current high-yielding ones. To avoid this disaster scenario, it is essential to have globally organized seed banks that carefully preserve the many different varieties. Several countries (including the UK) have made some attempts at storage sites, and there is a seed vault in Spitsbergen which has some 1.5 million types of seed samples. The location is in the Arctic Circle, so it will continue as a low-temperature store even without artificial refrigeration. Such stores include wild varieties (currently viewed as weeds), as this increases the genetic diversity of the store.

In terms of the economics, the monoculture large field is never going to be near the consumer, so real costs and efficiency must include transport costs, and perhaps refrigeration, as the products are moved to market. This is critical for perishable goods. Further, continued growth of the same crop will depend on the addition of fertilizers. Their usage can be reduced if skilfully applied at the correct point in the growth process, and also if they are targeted at the crop, not by some aeroplane spraying randomly over a field, with material blowing in unwanted directions.

However, added fertilizers become less effective over time—a 60 per cent drop in yield over ten years is not atypical. To an impartial observer, the arithmetic looks very different than it does to the farmer, who initially saw a big boost in productivity (profit). Because a 60 per cent drop in yield may be below even the original productivity, and certainly it will if the treatment costs are high, there may be a net fall in profit. This will be hidden for most people, as the change in prices with inflation over a ten-year period (typically a doubling) will disguise the fact that the real value has dropped. So overall there is a hidden economic factor that, just because the yield increases with fertilizers, there is no guarantee that the process is more economic. Once locked into the fertilizer and herbicide system, the cost of the chemicals may increase the production price far more than the yield. In agriculture, profit margins can be very small, and driven by supermarkets, rather than the public, and there is often a very fine line between profit and failure.

A major weakness of the fertilizers is that they do not stay on the original field, but instead are washed into the surrounding waterways where they are equally effective in promoting growth and new plants. This is a global problem, and various numbers are quoted from all the major countries implying that typically at least half of all water systems are contaminated by run-off from agriculture. This includes not only fertilizers but also insecticides, etc.

As I have emphasized, for large acreage of a single crop, there will not be natural predators to deal with pests, so there must either be a reliance on having a crop that in some way has been modified to resist a particular pest, or there must be spraying with insecticides. Historically, modified crops may have developed via natural selection, or more recently by an induced genetic modification. The latter is a highly emotive route, and it tends to polarize opinions of the various protagonists, but a rational assessment will probably find both positive and negative arguments with evidence from both viewpoints.

Farming attitudes and black grass

Plants called black grass emerge from the soil and grow in competition with many of our key crops (winter cereal, rapeseed, etc.), and in so doing reduce yields and contaminate the crops. They are particularly interesting because, although they have existed for a long time, they initially were successfully treated with herbicides that killed a large percentage of them. However, as in most plants, there are a minority of resistant strains, and these have expanded to become the dominant variant; without competition from the less resilient strains, they have become a major nuisance. The technocentric agricultural and chemical industries of course have tried to improve herbicides to win this battle, but are now struggling with a weed that is resistant, or susceptible only to chemicals that also kill the crops. So, further developments would involve a prolonged and extremely costly research programme, without a guaranteed, trouble-free end product.

In a very encouraging response to this difficulty, both the chemical companies and the EU Sustainable Use of Pesticides Directive have reappraised the problem and suggested a number of non-technological attacks. These include sowing more robust crops that outgrow the black grass; ploughing to bury the black grass seeds; changing the timing of planting and harvesting; and a culturally more dramatic reversion to crop rotations or leaving the fields fallow on a regular cycle. A mixture of these options appears to be successful, and does so with greatly reduced use of herbicides.

The perversity of these successful strategies is that they are not necessarily welcomed by farmers who have now become accustomed to finding a new magic herbicide, and want to continue farming in the same pattern. They have a dilemma, because many have moved to very large fields and invested in expensive machinery to cultivate them; furthermore, they are under contract to processors to grow a single crop and plant it at certain times of the year. Having heard discussions on the problem by farmers, it appears to me that they view reverting to a multi-crop rotation approach as being retrograde, not least because it was the pattern used by their grandparents. They therefore have a psychological difficulty in accepting that such a route was effective, and had been for many centuries. For the last 50 years they have been indoctrinated with the ideas of very large fields and lot and lots of chemicals.

Nevertheless, the growing pattern of reduced crop yields, contaminated products, and ever more expensive chemical treatments may well influence a change in attitude. There is also a possibility for some to form collective-type farming patterns where crop rotation is feasible within a group of farmers, hence the specialized sowing and harvesting equipment will not lie idle. Perhaps the most encouraging feature is that the change in attitude is also motivated by the agrichemical industry, which cannot see a rapid solution to the black grass problem. Once they—together with the farmers, buyers, and the public—realize that crop yields can be maintained with less chemical contamination and a better overview approach, there will be real hope for future food production.

A return to a diversity of breeds

My black grass example is about controlling a weed that is being selected into a resilient variety because of our prolonged use of chemicals. This is of course a very widespread problem, with many other examples. The obverse problem is when we have engineered a single crop strain because of some particularly desirable quality, then we are in equal danger of being hit by a disease that attacks this single strain; then the entire crop would fail (as with the Irish potato blight). Whilst monoculture may seem tempting in the short term, it is a highly dangerous strategy in the longer view. Many crops across the country are grown from seeds from a very limited range of suppliers. Therefore a major disease would not be confined to one region, but would blanket the country very rapidly.

The monoculture concept is not limited to crops—it has also been encouraged for many animal breeds. In the 1960s, the UK government was actively promoting the idea that only two or three breeds of cows, sheep, or pigs should be maintained, as these were high-yield varieties. At the time, there was some immediate logic, but it was extremely short-sighted, as many of the other breeds have qualities and resilience to disease, ability to survive in different conditions, or offer products that are beneficial for different dietary or medical conditions. Again common sense has prevailed to a limited extent, and the value of these less favoured, or rare, breeds has been appreciated and expanded in the last few decades. For me, the message is that there are benefits in a limited range of crops or animals in terms of mass production, but it is essential to preserve a far wider variety as an investment against future diseases and climatic changes. In non-financial terms, it is worthwhile to remind ourselves just how many species we have already brought to extinction and how difficult it is to reverse, or even reduce, such a trend.

Fishing

I have discussed farming and agriculture, but so far have not seriously looked at the problems of fishing. In part this is because most of us will visit the countryside and recognize the changes that are taking place. With fishing, however, the activity is not generally visible; nor can we see how well the shoals of fish and other maritime creatures are surviving. Nevertheless, detailed studies have been made for the last half-century of fish populations from all the major regions around the world. A typical pattern is that, relative to the 1950s, stocks dropped rapidly to about a quarter of their original values. In many cases, the numbers have stayed at this new low level because the quantities of fish are so low that they are becoming uneconomic for the industry. Falling to one quarter may not seem too dramatic. However, for cod in the North Atlantic, there are reliable data going back a further hundred years. Here the numbers are quite depressing, as the stocks were then perhaps 20 times higher than at present. Although mechanization and the use of sonar and satellite imagery may all be valuable technological advances for the industry in terms of finding the fish, the improved techniques are causing catastrophic declines in the populations of fish in our oceans.

Attempts at local and global legislation have been made, but are ignored by many nations. At the more local level, the rules still seem to favour larger fishing fleets, resulting in hardship for individual fishermen. This is still an area that needs better thought-out strategies for control, along with a parallel effort in fish breeding.

Technology of mutations

Mutations are a natural part of evolution. They are caused by chemicals and natural background radiation. We rarely stop to realize that we are continuously being bombarded by cosmic rays that rip through our cells and disrupt hundreds of thousands of them on a daily basis. Some cells recover, some die, and some reform in mutated variants. This is how evolution works, and it is inevitable. Higher doses of radiation or different chemicals just speed up the possibilities, but cell destruction and restructuring processes are ongoing. Our body has developed cells and chemical processes specifically to deal with repair and removal of damaged or altered cells, but with time there is accumulation of debris, and our recovery mechanisms weaken with age. So ageing and cancers (i.e. basically cells that are damaged and out of control) are inevitable—our only input is to influence the rate at which they happen.

Laboratory-induced genetic mutations may not be intrinsically different from natural ones. However, better understanding of DNA, chromosomes, and the building blocks of our cells has often revealed specific sites that influence particular characteristics. Genetic engineering can then target such sites. Ever since the Industrial Revolution, we have been bombarded with images and mental conditioning that engineering is the way forward—it implies progress. Therefore the phrase 'genetic engineering' is excellent marketing, as it has very positive emotive overtones. Nevertheless, we need to understand that Victorian advances in, say, steel making were broadly successful, but steel quality is still variable and failure can occur, different steels are only useful in certain circumstances, and we still have only a partial understanding of the details of metallurgy. When we had negligible understanding, we nevertheless had useful metallurgy (e.g. the early Bronze and Iron Ages).

Genetics is far more challenging than making better iron and steel for bridges. Living cells are complex: they can vary enormously and transmute into different structures. Further, the way they encode information is poorly understood overall, despite excellent identification of some key sites. Metallurgy has taken over 4,000 years to reach our current level of partial understanding, but we have only invested less than half a century so far into genetic engineering. Therefore we are unbelievably more ignorant than we wish to admit, and are deluded by the apparent speed of advance. This means we need to apply more care and caution with genetic engineering than we currently do.

Writing as a scientist, I recognize the danger that if we are trying to achieve a particular result, and we succeed, we will relax our caution and do not continue to look for related, unwanted effects. There is a purely practical reason: the funding for the research will invariably have stopped. But in the case of genetic mutations, we need to continue to be vigilant, looking for additional changes in the cell and plant (or animal) behaviour for a very long time.

Evolution and mutations are emotive terms, but they exist naturally, and for thousands of years we have been exploiting them under the different name of selective breeding. Cattle, horses, and dogs are all familiar examples of dramatic changes from natural-source animals. Dogs, for example, seem to have a common genetic ancestry in a small group of wolves, but we have interfered; thus modern examples range from Great Danes to chihuahuas, bulldogs, and Daschunds. For rapid-breeding creatures and plants, the changes can be produced quite quickly. If we tried the same approach with humans, it would be far more difficult to control over many generations.

Nevertheless, many of the breeds we have 'designed' have weaknesses (no matter how attractive or useful the animal). Dog breeds have a whole range of faults, such as deafness, short life expectancy, tendency to cancers, hip dysplasia, heart conditions, problems with whelping, or poor breeding. These are familiar to the breeders, and can be informative in the sense that they often reveal genetic links between, say, deafness and patches of different coloured hair (also seen in humans), which are valuable data for geneticists. Compared with naturally evolved wolves, most dogs we have engineered are inferior specimens that could not survive without our veterinary support.

Our more extreme experiments of cross-breeding different species suggest that in those cases, one generally does not have to worry about longer-term, inherited mutations, as the progeny are often sterile. A familiar example is the mule, which has excellent characteristics for a pack animal in transport, but is a one-off, end-of-the-line species. To some extent, the same problems have existed in many cross-bred plants. Overall, it is unwise to be too specific in attempts to develop new hybrid species; my guess is that it is a topic that will develop as knowledge and skills of genetic engineering advance.

In agriculture, many crop variants that are effective and widely used are sterile. So the farmer cannot retain seed from one year to sow the subsequent crop. This example of sterility may have been added intentionally during production of the seed, as it forces the farmer into buying more seed each year. I view this manipulation of the farmer to become a pawn in the hands of the agrichemical industry as iniquitous.

Evolution also applies to the pest and diseases we are trying to battle; a chemical that is successful will never be so permanently, because the pest survivors will expand to take over from the variants that were destroyed. This 'bounceback' resurgence is well documented, in part because the treatments kill the natural predators, as they die of starvation. Just as with medical usage of drugs, new strains are often harder to counteract than the original ones. The battle between biochemist and bug is increasingly demanding and costly. Reliance on natural bug controls is therefore the optimal solution.

However, the difficulties are compounded because of global trade: new bugs and plants are continuously being imported into all regions. Because they rarely arrive with their natural predators, they thrive in their new homes. This is not a new problem. Even in 1962, Rachel Carson noted that in the USA, some quarter of a million non-native species had been identified. In the subsequent half-century, the number is likely to have doubled.

Positive examples have been demonstrated in which the predators of foreign species (whether bug or weed) have been imported to attack the foreign pest. In the best scenarios, the second import only thrives off the pest, so does not contaminate the rest of the native population of plants or insects. Nevertheless, this is a tricky solution to consider, because if the climates of the original and new region are different, then either pest or predator may adapt differently.

Water

Water is the key to every form of life, so it is worth looking to see how changing technologies influence the availability and quality of the water needed for agriculture (and humans). Where water exists, plants can grow and animals graze. The effects are often spectacular: to see a beautiful green golf course in the middle of a desert is an instant reminder of what it can achieve.

Free delivery by rainfall, in the appropriate amounts and correct timing, is a farmer's dream, but more typically it is variable, and irrigation must be provided. This requires construction of dams, aqueducts, canals, and wells or containers, so is costly and labour-intensive. Once delivered, water will soak into the soil, run off with the fertilizers, evaporate, or be absorbed into the plants and shipped away with them. In each case, the net effect is a loss of water that needs replacement. We are now using water from ancient underground stores at rates that mean it cannot be replenished at the speed we extract it. So the water we are now drinking may have come from rainfall a few thousand years ago, but the volume available is shrinking.

Water extraction also causes the land to sink. In some areas this is quite obvious with examples in, say, California, where there is extensive water pumping technology, and in some isolated areas the ground level is dropping at rates as high as a metre per year. The problem is widespread; over the last century, entire areas have dropped as much as 10 metres (~30 feet).

Water is also used in mining, oil extraction, a vast range of technological processes, plus home usage and sewage systems. The pattern in each case is that we may have started with a pure water source, but the processing always degrades it. We may choose to live with the contaminated water and reuse it, but perhaps a disquieting thought is that the water drunk from the taps in London is claimed to have already passed through seven other people on the way there. They have added to the flavour with traces of their food, chemicals, and drugs. Sometimes we are convinced we can taste the history of the water, but generally these are just from the chemicals used in the attempts at purification.

Nevertheless, modern analysis, which can achieve sensitivity down to parts per billion, will definitely reveal traces of drugs (of all types). My earlier example was drug detection in the Italian river Po, but this is universal, and London can currently claim to have the dubious position of having the highest cocaine content in the water supply of any major city. Our concern is that water supplies inevitably contain detectable traces of drugs, impurities, and a wide range of biological compounds linked to diseases. Switching to wine can dull our concerns, but it does not solve the problem, as wines also have the chemical signatures of the water from their local regions.

For the high water volumes needed in agriculture, not all used water can be adequately recycled; therefore large quantities become unsuitable both for human consumption and for irrigation. The story does not end there, as polluted water heads down the rivers and out to sea where it interferes with, and contaminates, marine creatures and fish. They may even detect the pollutants better than we do because as I already mentioned, many species, such as salmon and turtles, return to their breeding grounds by the 'chemical smell' of their home rivers or beaches, so they certainly are actively sensitive to the parts per billion of contaminants.

My historic examples of collapses of civilizations from drought should keep us aware of our dependence on water, and force us to treat it with care and respect. It may fall from the heavens, but it is not guaranteed. Anyone can contaminate it, but purification is difficult and very expensive.

Optimism or pessimism?

I have deliberately tried to focus on issues raised by Rachel Carson back in the 1960s, as at that time, her writing and understanding of the downside of farming practices were groundbreaking. She raised awareness of key issues of how we are destroying our environment, both consciously and unconsciously. The problems are compounded by ignorance, by being blinded by scientific advances we do not properly understand, and by short-term profits and higher crop yields. Her message was clearly delivered, and in that sense she was successful. However, over the same half-century, the world population has doubled, and our expectations of better lifestyles, more food, and more exotic input to our diets have soared. So the problems of producing more food are actually increasing, and despite far better understanding and knowledge, they have not gone away.

The only way that there can be genuine improvements is to encourage, or perhaps ensure, that population growth is slowed and reduced globally. Any imposed mechanisms to do this will be highly contentious. Perversely, we might benefit from a major global disaster such as the plagues and diseases that have periodically hit us. There is no doubt they will occur, and they may even be variants of earlier ones such as the Black Death or the 1918 influenza epidemic, both of which brought major loss of life to Europe. In those examples, the localized populations were reduced by up to a third in some areas. In a more widely connected world, such epidemics will not be localized in a single continent, so population and economic impacts will be even greater.

Recognition that we are also wasting a vast quantity of the food we produce, and many nations are overeating by excessive amounts, could additionally buy us some breathing space in terms of food production. Cutting overeating would have the additional benefit that it would generate a far healthier population.

If—and it is an incredibly big if—we can reduce and stabilize the world population, then the planet could support us and simultaneously preserve the environment and other creatures that have an equal right to survive. I find it distressing to admit to myself that I do not believe we will manage this. In part, this is because our numbers are soaring; less obviously, it is because so many people exist in cities, and consequently have no concept of, or interest in, the wider world of agriculture or the natural environment. For them, these are just interesting items to amuse them on the TV. Their real connection to food production, farming, and other areas of the world are rarely much different from watching science fiction, historic fiction, or crime series and soaps.

Whilst most aspects of this book are about the dark side of technological innovations, I suspect that the truly catastrophic potential of global exploitation and destruction is primarily unrelated to technology, and related instead to the expansion of population, as well as to self-interest and human nature. Technologies are just the enabling routes to self-destruction, not the cause.

8

Medicine—Expectations and Reality

Medicine—the scale of the problem

When I started to plan this chapter on the dark side of medical practice, treatments, and drugs, I assumed that my problem would be to select the most entertaining examples, or the most horrendous. My view was conditioned by a vast number of anecdotes, experiences of friends, and examples from the numerous stories in the media, as well as the Internet, articles, and books.

These are frequently not just the views of outsiders, but include equally highly detailed and entertaining books, such as those of Ben Goldacre (*Bad Science*) and Robert Winston (*Bad Ideas?*), both of whom are successful medical practitioners. In fact, these insiders' views are revealing, and both authors are equally critical of past and current practices.

Therefore my first impression was that there are some aspects of medical practices that are seriously wrong. This is based on the wealth of examples of failures, incompetence, or unfortunate side effects that are being quoted. It is a vast field, so inevitably there are examples that fit this preconception. It is, however, only partly true, and I have tried to avoid falling into this simplistic trap, but instead have tried to rationalize why there are so many examples of failure (and, of course, success).

The essential realization is that I am not dealing with the simplicity of faults in Victorian plumbing, or some limited product from a few hundred producers, who may be entirely conscientious and hard-working, but have not predicted or expected a long-term side effect from their pride and joy. Instead I am trying to focus on the health and well-being of humans, and this seems to concern us so much that it was, and still is, an incredibly dynamic and competitive business. The sums of money involved worldwide are many billions, so I am not surprised it attracts not only highly intelligent, dedicated, and able people, but also those driven by ego and self-importance, as well as many who see it as a career with an easy source of income, rather than one for the benefit of humanity.

For us, the poor customers and patients, it is incredibly difficult to separate out the true and different qualities of experts, and the validity of claims made whilst marketing the products. Hence, we will eventually have many disappointments and bad side effects, as our expectations are too high. Rather worse for our public understanding is that medical opinion and knowledge are highly dynamic. New ideas and results emerge on a daily basis, so there are often bitter conflicts and public arguments as to the value of treatment and products. This is extremely confusing, as often the conclusions are based on precisely the same data, but the presentations are driven by egos and status. (This is not just a medical problem but is typical of many leading figures in all walks of life.)

In order to try to gain some appreciation of the scale of this industry, I can cite the current number of employees by the National Health Service in the UK. This is around 1.3 million employees, of whom nearly half a million are qualified doctors, nurses, and dentists, plus all the ancillary staff. This is in a country of around 56 million people. If we add in those employed in the pharmaceutical industries, those working in biological sciences, and people running private practices, a sensible guess at the overall total of skilled and qualified workers is nearer one million 'experts'; i.e. around 3–4 per cent of the working population have different types of expertise in medicine and biology. It is high, but it is not an atypical percentage for advanced countries. For example, the US 2015 listing of professionally active physicians is almost 1 million. Inclusion of the other skilled support staff gives a similar percentage to the UK.

Such skilled people will be a smaller percentage in poorer or underdeveloped countries. So for the current 7 billion inhabitants of the planet, we can make a reasonable estimate that there are maybe 80 million people worldwide who influence our health. This is excellent news in terms of the probability of making progress and gaining a better understanding, but the sheer scale of the numbers means the literature and ideas they produce can be totally lost within the immense volume of data, knowledge, and statistics they generate.

Also, if just 1.25 per cent are incompetent, or charlatans, then we have bad input from more than a million people! Realistically, I am sure this will be an underestimate. Overall, with such a volume of effort, the presence of a serious side effect for some drug, or medical concept, may be totally buried in the literature, and be unread by most of the experts and the public. So rather than view all the problems we do know about as a dark side of medicine, most examples will still fall in the category of ignorance and inadequate dissemination of the facts.

We may also assume that we have a much higher percentage of the population involved in medicine than in the past, but realistically, even in small tribal communities, there will have been (or are) herbalists or witch doctors. Their knowledge and effectiveness may be less than that of modern doctors, but as a percentage of the community, they may well be a similar fraction of their communities as we have in more developed societies. We may not appreciate that in more advanced countries, there were also herbalists, apothecaries, and barber and battlefield surgeons, in addition to the more formal medical practitioners.

Attitudes and expectations from experts and the public

Despite the hype and publicity, our expectations of medicine are unfounded, as medicine can never be such an exact and 'hard' science as, say, physics or chemistry. Physical principles are the same everywhere. For example, whilst we may have computers that do not always function as expected, the failures are invariably caused by the human-written software or bad circuit design—the underlying electronics of the computer itself are totally predictable, once we have understood them.

By contrast, all 7 billion of us are different. We differ right from birth in terms of our genetics (even for twins), and hence in our responses to environment, nutrition, climate, and lifestyles. We are influenced by our types of work, as well as by our leisure pursuits. We are equally unique in terms of the exposure to diseases and our responses to them. Such differences are cumulative, as are long-term effects of medical drugs, diet, alcohol, and all other substances that people consume for pleasure, and indeed all other aspects of our working life. Hence treatments, or overall attempts to improve the quality of life, are far more challenging than, say, the problems previously discussed in finding herbicides to modify or improve crops.

Nevertheless, in the case of drug treatments, there is the same underlying problem that development of drugs to attack a specific disease will never be totally successful for all people. Some patients may not respond for genetic reasons; others—because of their specific medical history or because of other treatments they are undergoing—may have unexpected side effects. This means that even for well-established drugs (such as aspirin or penicillin) that are normally star performers, there will be failures or complications.

The downside here is that our expectations are too high; there will never be universally successful treatments. Websites that detail such failures may focus on the fact that safety testing is invariably made on animals; the sites are often making the perfectly valid criticism that it is cruel to the test animals. Further, the websites indicate that upwards of 90 per cent of drugs that pass animal tests then have failures with humans. The logical conclusion is that the tests should have been made on humans—but the ethics of this are equally flawed. Examples where it has been done in the past, often without the knowledge of the test subjects, are now rightly condemned. Safety testing is therefore an ongoing necessity, which will never be anything like 100 per cent successful. Basically, this is an insoluble problem, and we must just accept our best efforts.

The variability of human response means that virtually every successful and widely used drug comes with a warning label that there can be numerous side effects. This is true of even the most minor and commonly used prescriptions and over-the-counter pills. In terms of numbers, from regions with reliable statistics, we can quote that in the USA there are annually around 4.5 million visits to surgeries or emergency caused by adverse side effects to prescriptions. Perhaps even more surprising is that, in addition, there are a further 2 million problems with patients already in hospital.

The test problems in agriculture for herbicides and new strains of crops (but not on the longer-term movement of chemicals to other environments and people) are much simpler than for medicine. For herbicides and engineered crops, we may make mistakes, but to a large extent we have control of the growth conditions and the environment. Plants for food, such as wheat and other cereals, have short growth cycles of just a few months. Further, we are not concerned if the experimental plants are destroyed. For crops, there is a bonus of very large trial populations, so we can produce good statistics on our interference in normal plant growth (e.g. for wheat there may be 5,000 plants per acre). The other factor is that plants may be living entities, but they are not sentient, and we have no emotive view of them. Perhaps an additional, but a slightly cynical comment, is that if our experimental interference in their life cycles goes wrong, they cannot sue us for negligence.

An additional feature is that experiments with crops invariably have limited objectives (e.g. combating a particular pest or disease), and we can use different herbicides, or growth factors, either separately or in combination. This is not possible to the same extent with trials with human patients, where not only must we cure the initial problem, but are concerned with long-term health issues that might develop as a result of our treatments.

The unfortunate, and often disguised, reality of the developments with crops is that it is obvious that within a short period of time that most herbicides become ineffective. This is partly because some of the pests and diseases are immune to our treatments. The same is true of all the key drugs that we use on humans and animals, and now many strains of disease have genetic survivors that are resistant to antibiotics. We often fail to recognize this. Our expectations over the last 50 years were unjustifiably excessive, as we had the typical human response that there would be a totally successful solution to our problems via new drugs and medicines. Using higher doses or variant drugs is not a solution; it is counterproductive.

Developments of genetic resistance to treatments is just part of the natural selection of evolution, so this should no longer surprise us, as our modern biological skills mean we can even sequence the genetic codes to recognize differences. In many cases, the variations are valuable for humans. The classic example of this is that some people, and their families, were immune, or survived, the Black Death. The reason is now clear, as modern analysis shows their descendants have a genetic mutation that offers immunity. Similar examples are now known for other killer diseases. Indeed, without such genetic variations that offer survival to a limited fraction of the population, the human race would have been exterminated long ago.

The DNA sequencing data, and the claims made from it, should also come with a health warning. Although we can frequently detect differences between people, we still only understand a tiny fraction of the genetic code, and certainly have minimal detailed knowledge about cooperative effects between different regions of our DNA. So simplistic meddling to 'improve' the human race is currently far beyond our predictive power, except in a few examples of genetic errors that generate specific diseases. Some genetic treatments exist, and they are fantastic, but we must not expect equal success in every case.

Understanding side effects and drug testing

In treatments and experiments with 7 billion variants of humanity, our best hope is to have successful medicine for a majority, and expect errors and side effects from our efforts. We definitely have no hope of always making the right treatments, or, more crucially, actually fully understanding what we are doing. So rather than only consider living subjects, it is worth briefly looking at inanimate parallels.

I am a professional physicist, and during my career have quite often encountered conflicting ideas and results from different research groups, or industrial processes that were far from optimal. In some cases, I personally have made, and eventually realized, that I and the other groups across the world have been making systematic errors or serious misinterpretations of data. This is initially embarrassing, but it is fine—it is life, and it is progress. There are of course difficulties in getting the new ideas and corrections published, but with delicacy and non-emotive presentation, it has always been successful. Some people have of course been reluctant to admit they also were wrong, but within a couple of years most have changed their habits, and overall our field of work is slightly more reliable.

By contrast, the medical literature—which is almost certainly dealing with more difficult and less precise situations—seems to become far more polarized if different sets of results, or safety studies, are in disagreement. There appears to be a determination to oppose new data or change long-term practices. The discussions with criticism often seem to be highly personal, and far more vehement than in the 'hard sciences'. The reason for this difference in attitude seems very strange, as for medicine, errors involve people, the patients, and we need to consider their health and survival. I will leave any readers interested in this oddity to explain the difference in attitude (it may not involve any commercial interests, just reputations).

Nevertheless, for us, the public, it becomes incredibly difficult to assess what is our best practice in terms of the treatments we are prescribed and the drugs we take. Access to the Internet is helpful, but again published items will include bias, and opinions that change with time as new studies are made and evaluated. But, at least the pattern of doubts and errors may possibly become apparent.

I am justified in being extremely cautious, as one can find highly respected overview studies that make precisely the same point. For example, I saw one that claims of the 50,000 or so new journal articles published each year for clinicians, the pattern is that they, the reviewers, believe no more than around 6 per cent of the studies are well designed, relevant, and unbiased! They even suggest roughly $200 billion is wasted on poorly designed or repetitive studies.

The difficulties do not stop here. In their overview for the public, they use an example of whether certain foods reduce or increase the risk of cancer. Items range from wine and tomatoes to butter and beef. Results are *very* divergent, and they offer a simple diagram to show this. This has data from various studies indicated by quite different points along the positive or negative axes, indicating the spread of opinion from good to bad. Fine, I do not need any scientific skill to understand such a spread in opinion. Most of us will look at the diagram and ask, 'Where is the average value?'

Beware! Only with more care did I notice the graph was plotted with logarithmic axes, labelled in small print. I doubt that the general public will notice this, and if they do, they will be totally confused as to where the average value occurs. Our normal response is that if we cannot understand something, or if it disagrees with our preconceptions, then we will ignore it. This is particularly true if it means we should make a change in our lifestyle.

With the variations between people, it is inevitable that many of us will respond unfavourably to some medications that are highly successful for others. This immediately shows up in the more widely used items. For example, one extremely familiar pain killer has global sales of around 100 billion pills per year and is available over the counter. On this scale, it would be unbelievable if negative side effects had not been reported, and indeed the manufacturers list more than six types of allergic reactions, more than six serious side effects, and a list of other minor factors. The list includes gastrointestinal issues such as internal bleeding, nausea, constipation, and diarrhoea induced by the drug as it passes through the digestive system. Other side effects listed are drowsiness, pain, skin reactions, and further difficulties with extended usage such as stomach ulcers. These effects are unpleasant. Nevertheless, if one has a bad reaction to the pills, then it is easy to change to an alternative in the future. Overall, the scale of the side effects is minor compared with the usage, so is acceptable.

Alternatives to this particular item are also widely available; in the UK, some 30 million packs are sold each year, and, of course, even for the recommended dosage, some users will suffer. In this alternative, exceeding the recommended dosage can be lethal, and this is clearly stated in the packaging. For both items, the benefits for the majority outweigh the inconveniences for the minority.

A far more complex situation exists for prescription drugs, so I will mention one example where the Web literature is strongly worded from different perspectives. This is for the widely used drugs called beta blockers that are prescribed in the treatment of high blood pressure and congestive heart failure. Apparently, in this role they are effective and can work successfully. As a result of an initial study, an eminent European cardiology body had also recommended them far more widely in non-cardiac situations. Side effects are well documented and include the fact that beta blockers can commonly result in dizziness, tiredness, blurred vision, cold hands and feet, slow heartbeat, diarrhoea, nausea, impotence, etc.

Despite the initial enthusiasm, and apparently positive responses, we now have hindsight, and a lot more data, and the downsides and the range of uses where they are undesirable, means the original view on them now seems to have been an error. More recent articles are bluntly claiming that UK doctors have been causing around 10,000 deaths per year, and across Europe the total may be 800,000 related deaths over the last five years! With the new perspective, data, and understanding, there are revised EC guidelines and legislation.

Unfortunately, and quite typically, there is a very considerable time lag of two to five years as general practitioners and hospitals move away from their familiar prescription habits and follow the new guidelines. Rather more unfortunate is that many doctors are overwhelmed with conflicting data on the drugs they prescribe, so that they never update and respond to new recommendations. The consequence is that, as for beta blockers, their prescription habits have not changed at all.

The medical behaviour and timescales are identical with my example of physicists only slowly recognizing they were making mistakes, and the reluctance to change is just part of human nature. However, for medical errors, the consequences are far more serious.

In a later context, I will also mention the conflicting views on the use of hormone replacement therapy, which has similarly attracted heated and prejudiced attitudes from 'experts'. The conflicts there are interesting as often they use the same data to draw their strongly worded opinions. An identical spread in views exists for statins; many websites report studies implicating them in a range of medical problems (including Type 2 diabetes), but these studies are rejected by those with entrenched views, who instead are saying they should be taken more widely on the basis of age, not need.

Do we need such an immense medical system?

Until around 150 years ago, medicine and the understanding of the human body was not very advanced. In the first few thousand years of our history, medicines and drug treatments relied on good fortune in finding natural herbs or other substances that had positive effects to offset illness. The herbal and early medical skills were gained totally by trial and error, without understanding the chemistry and biology of the treatments. Survival and success might just have been by good fortune. Successful treatments have existed; an oft-cited example of very early surgery are the skulls that show that trepanning was done with Stone Age sharp flints to relieve pressure on the brain. The scars on the skulls prove some patients survived. That was impressive.

Nevertheless, there was negligible knowledge of how the body functions, or how diseases were transmitted. Surgery was horrendous, with most of the skills initially learnt by battlefield, or gladiator-type combat, repairs. The lack of anaesthetics meant speed was often more critical than delicacy and precision. Invasive actions on the body are perilous, and particularly so in earlier times as the surgeons had no concept of the importance of cleanliness and sterility of their tools and working conditions. Ignorance also meant that the introduction of simple anaesthetics actually increased the infection rates. Because there was more time to operate, there was a parallel increase in the exposure time for infections. If we look at Victorian pictures of surgeons, in ordinary clothing, with tools that were not washed between patients, it is amazing to realize that they were so ignorant of the way diseases were transmitted. Indeed, many surgical tools had elegant wooden handles, so sterilization was impossible. Older doctors will admit that many of the general attitudes had not changed by the 1960s: the daily inspection of patients by the hospital supervisors involved no handwashing between patients, and allowed no comments or questioning in procedures from the lower ranks. Some medical friends say this is still the case.

In the community, the local doctors were equally omnipotent, and neither could they be challenged by the patients, nor was such a possibility considered. This hierarchical structure had some benefits in that hospitals were controlled by those with medical knowledge who understood what was required, whereas the current impression is that the managers are only interested in balancing the finances, independent of the medical priorities. There is also now more effort on image—to have top surgeons, or the most expensive equipment for analysis and diagnosis. This change in pattern certainly is included in my list of the downsides of technological progress, as all too often hospitals lack the money to fully exploit their expensive toys.

The pattern of effectively working a five-day week in some sections of UK hospitals means the expensive equipment and top skills are often unavailable on weekends. Investment in people would open up equipment use on a seven-day-a-week basis (i.e. a 40 per cent gain in efficiency!). It needs more staff, but not more major equipment, which makes the suggestion economically attractive. Far more important, it would improve the survival statistics. These are already well documented, with a clear negative weekend admission effect. Numbers vary between hospitals, but various studies across the UK are relatively similar (as of 2016). They reveal a 20 per cent higher risk of dying for stroke patients who are admitted at the weekends, and overall a 10 per cent lower survival rate of all admissions. To offer a number scale in terms of people, the estimates for London are 500 deaths per year attributable to the current inadequate weekend hospital treatment and skilled coverage.

The blogs and Facebook comments from the medical profession saying they are already working a seven-day week only add support to my comments. They often include information that, in order to have qualified weekend staff, they are working 24 hours or more at a stretch. They may be skilled and dedicated, but there is no way they are superhuman, and I would not wish to be treated by anyone who has worked such a long shift. Human concentration falls dramatically over the course of a few hours, and our error rates soar over such a long period of activity. I suspect these extreme working conditions are actually increasing the weekend failure rates.

The solution is more people, not longer hours. In countries where seven-day coverage exists, and long working days are banned, there are no anomalies such as the UK weekend slump in survival. Nor do they have the difficulties, as exist in a local major hospital, where the weekend delay in being treated in accident and emergency is currently averaging a four-hour wait (yes—I wrote four!).

Personal contacts between patients and doctors

Another change that has happened as treatment has become accessible to everyone is that the doctors often have so many patients, or are part of a team of doctors, that they no longer have any close knowledge of their patients. They may not even recognize who they are. For example, in local surgeries, the occasional visitor will rarely see the same doctor, and at most is likely to be in the surgery for less than ten minutes. This is the target number. Diagnosis is inevitably very rapid, and it puts intense pressure on the doctors to perform consistently. This retrograde step is mainly driven by financial constraints and administrative procedures. Additionally, demands on analysis technologies have increased and are made elsewhere. In principle, this is sensible, but one hears many stories where the results are not fed back to the patient. Less often, the central analysis sites are overloaded and records are confused.

A further downside of the multi-doctor practice is that the patient may be seen by those who have a very limited record of the patient's history or lifestyle, and consequently are unaware of a patient's responses to different drugs. This may not be serious for routine problems, but it can (and does) lead to errors for more complex conditions, (e.g. where the patient has serious allergic reactions to certain drugs, or even intolerance to the lactose content used in the drug fillings).

The official response to such criticism is that there are computer records and files, so this knowledge should be available. However, from the experience of my own friends, this is certainly not the case. For example, one man with a rather long record of treatments had a hospital file many centimetres thick, but rarely was seen by the same members of staff. There is no way the latest expert would have had time to read all the background history, and only intervention by a determined wife stopped him from being given drugs to which he had already shown major allergies. Unfortunately, this is not an atypical example.

General practitioners need an exceptionally wide perspective and breadth of knowledge, but such breadth is only feasible at a non-specialist level, so one cannot expect perfect diagnoses for more exotic conditions. In this instance, modern technology offers some advantages, as a cautious Web search of conditions that match one's symptoms may actually offer more detail and diagnosis than one could reasonably expect from a doctor who is seeing a different patient every ten minutes.

I also have a personal criticism of the apparent attitudes of the medical system in that there is an emphasis on trying to use drugs, surgery, and all the other medical advances to treat patients to prolong life. For me this is often totally wrong, and the priorities should be to maintain the quality of life, not the length. I have had a number of friends and relatives who have said that the treatments of terminal conditions were not worth the extra few months that it might have achieved. In some cases, they consciously stopped treatment because of this. This is not an unusual situation, but it needs determination and family support.

Gullibility and marketing

Rather than be critical of the medical profession, which is struggling against a massive challenge, I suspect that the real culprit in terms of the cost of the health system is the public itself. Throughout history, we have been incredibly, and willingly, gullible about the efficacy of drugs and treatments. We have always wanted some 'magic potion' that would fix our problems. It might be the 'elixir of love' that makes us attractive and irresistible, or just something to stop us ageing or going bald. The only difference is that the pedlars of such rubbish were once only found in the local markets, whereas now they are in glossy advertisements and on TV, or use the Internet and mass-market junk emails. The range of products is imaginative (Ben Goldacre's book *Bad Science* offers very many different examples). In general, we are readily deluded, not just by the products and presenters, but because we want to believe in a magic cure, no matter what is the evidence or common sense. The historic examples of street market pedlars have merely shifted to the new media. The marketing is of necessity more polished and more persuasive. The glamorous presenters in white (or blue) lab coats, which add gravitas and imply medical knowledge, make stupendous claims for their products. Ideally, they look super fit with luxuriant hair, glowing skin, and shiny teeth, which we somehow imagine are the result of using this particular product. In addition, the advertising claims of scientifically tested and proven products can be highly dubious, especially if the testing was done, or sponsored, by the manufacturer. In many cases, truly independent testing will reveal minimal benefits. But we, the willing victims, want to believe, so we buy the goods.

Another key element of this progress in marketing psychology is that, even if the claims are extreme, they will sell if marketed at a sufficiently high price, and advertised on TV by glamorous popular film and media personalities. We apparently routinely assume higher price implies a better product, and we trust those we subconsciously identify with. TV advertising can equally be unhelpful for our health where there are advertisements that basically say, Sit back, relax, do not worry, eat our product, instead of saying, Go out and do some exercise and eat wisely.

These are mostly errors from our imagination, as we see what we would like to believe. For example, a number of TV and radio experts have happily implied they have medical knowledge and training, or degrees in medicine. Reality is of course different, but virtually no one will ever check on this. Indeed, there have been examples of hospitals employing new 'doctors' who have had zero formal training, but no one has followed up on their references! More worrying is that there are examples of such 'doctors' who have functioned successfully for many years.

Just to emphasize that carelessness in checking records and details is not confined to medicine, I can cite a personal example. On one occasion, I received a letter offering me a distinguished and well-funded sabbatical in a foreign university, in a subject totally unrelated to physics. It said it was for world-renowned scholars (vanity obviously agreed with this), so it was tempting. Unfortunately, I realized they had confused two universities (Sussex and Essex) and they wanted a man of the same name, but from the other one. I also often had other items of correspondence, and occasionally cheques, sent to me! He clearly was leading a far more exciting life than I was.

In this section on marketing, it is worth noting that marketing techniques are not just aimed at the public but also at the medical profession. For the industries, persuading doctors to prescribe particular drugs or treatments is an excellent business strategy. Because immense sums of money are involved, the methods used may be more aggressive and persuasive than are justified. Whilst not illegal, encouragement and funding to attend conferences, etc. related to specific products is often effective in terms of an increase in prescriptions. Nevertheless, there can be conflicts of interest between organizers of such events and the roles they have within the national health industry. I am also aware that in at least one EU country the doctors receive a significant payback from the drug companies for the money patients spend on prescriptions. The quality of the doctors is not in question, but the psychological pressure to overprescribe certainly exists.

One may be just as critical of a recent development in the UK, where doctors receive £55 (~$70) for each Alzheimer patient that they diagnose. This seems totally unacceptable as diagnosis is part of the profession. Further, the typical GP has an income of nearly three times the average wage in the country. With a quarter of a million new cases per year (i.e. the population of a large town), the UK is wasting enough funds to have supported around 500 nurses per year. Registered new Alzheimer cases in the USA are currently at half a million per year. Relative to the overall population, this is less that the UK rate. However, the USA does not have a free health service, so fewer people are being counted. The US Alzheimer statistics record that more than one in ten of the over-65s are diagnosed with the disease.

Self-destruction

A far more serious weakness of the public, and an immense cost to the health service, comes from our own inability to take care of ourselves. I will pick just three types of example to show how a very large number of people are actively, consciously destroying their own health, and apparently do not care, not least as they assume someone else will eventually look after them once they fall ill (as they most certainly will).

The most obvious example is smoking. There are government health warnings on UK cigarette packets. They very clearly say that half the people who smoke will die from diseases related to smoking. This is not some new discovery, but has been in the medical literature and the public domain for a very long time (examples have existed for several centuries). Smokers will of course claim that they have the freedom to do as they wish, and are paying taxes on their tobacco products which, hopefully, contribute to the health system.

This is totally misleading as the taxes do not take care of the inflation that occurs by the time the smokers need treatments, nor do they provide sufficient new hospital facilities and staff. Mere tax money cannot cover the losses and disruptions to their families who have to look after them. It would be interesting to consider if smoking would continue at the same level if treatment of smoking-related diseases were only maintained in private healthcare. This is not an original idea, as in many nations the only healthcare is for those willing and able to pay.

Publicity and education have helped to some extent, as the percentage of the UK population that smokes has halved since 1974, and the total number of smokers has dropped. Nevertheless, smoking still accounts for about 33 per cent of respiratory deaths and 25 per cent of those who die from cancer (plus featuring in many other types of illness). Here I have cited statistics merely as percentages, and for most of us this has little emotive impact.

Therefore, I will rephrase it and say that around 100,000 (yes, 100,000) friends, relatives, and colleagues are killed by smoking each and every year in the UK. This is equivalent to a medium-size town; in terms of suffering to them and their families, it is immense. In monetary terms, it is costing the country many billions. In terms of trauma, loss, distress, and the need to care for the people associated with the diseases, the annual numbers are far worse, since the level of serious illness may typically extend over a five-year period. From this overview perspective, the number of people caused to suffer, directly and indirectly, involves close friends and family, and over a five-year period, the number of other people with lives degraded by the smokers rises to far more than a million of us at any one time. This is just in the UK.

Better health education with a much more aggressive and focussed objective is needed to really switch people's behaviour, not just increasing the health service budget. Education is cheaper than medical treatment. The oddity is that if 50 per cent of people who smoke will die from smoking-related conditions, why do people not realize that they are at a major risk? This is a far higher risk than virtually any other activity that we encounter in our lives.

Further, the evidence is very clear that poor breathing, etc. will also lower their quality of life long before obvious diseases set in. Consider as well the fact that people suffering the consequences will invariably be saying 'I wish I had not smoked'. This is not a problem limited to just the smoker, but it impacts on their family and friends. Help, support, and motivation to enable people to quit smoking are not just social ideology, but directly represent financial savings to the entire nation, and especially via the need for medical care. Savings, on a scale of many billions per year, whilst simultaneously reducing the strain on the medical services, are achievable realistic numbers.

The second very obvious self-inflicted set of problems comes from obesity. For the health service and cost to the UK, some estimates suggest the number is around £6.5 billion per year. Although there are genuine medical conditions that cause obesity, including inherited genetic ones, the majority of cases of obesity are self-inflicted by excessive eating and lack of exercise. The rapid rise in obesity in children is certainly compounded by their being transported to school (instead of walking), reduction in school sports facilities, and the fact that so many children are overweight that there is no peer pressure to make an effort to change. The euphemism 'obesity' also implies it is a medical condition for which they are not responsible. TV and media images further imply being overweight is acceptable in terms of image, and associated illnesses receive minimal discussion.

The problem is compounded by the fact that there is a vast amount of advertising pressure to buy more food, and there are opportunities to overeat, not only nutritious food and drink, but also a plethora of junk items. These may increase waistlines and volume, but offer little true nutritional value. Similarly, products may contain substances and flavour enhancers that effectively make them addictive. If they are marketed via sports personalities, they are automatically assumed to be good for health.

Labels on the products listing the sugar, fat, or caffeine contents, etc. are probably ignored once they have been added to the list of preferred items. Junk food can appear in unexpected ways; one such is in the use of cheese substitutes on pizza. Alternative products that look and taste like cheese exist. Nutritionally, they are of minimal value, but are cheaper than cheese. The financial temptation to use them in pizza outlets means they are widely (and legally) used. Overall, the proliferation of snack and junk food items, which can become addictive, is a threat to those who wish not to be obese.

It is easy to see evidence for increasing numbers of overweight people—if we look at films and newsreels from times as recent as, say, 30 years ago, then in the UK one saw relatively few fat people, and almost no overweight children. Over this period, there has been at least a threefold increase in terms of numbers of people who are technically obese. The problem is actually far worse when assessed in terms of just how obese many people have become. Also, the presence of so many overweight people means they no longer stand out from the crowd, and so do not care about their bodies. The changes over the last few decades are self-induced, as our population is genetically the same race, so the change and increase must primarily, and almost entirely, be blamed on us. Advertising of foods is not the sole problem as obesity is not totally uniform across the UK, with the worst areas being in the north-east of England and the west Midlands. Overall, the numbers run at a quarter of adults, and, absolutely shocking, is the rise to around one in six children. This is therefore a key problem, not just for those who are obese, but for the entire nation, which allows and encourages this excess to appear in our children. The young are not just being deprived of long-term good health, but also all the pleasures of fitness and activity that we should expect as the norm. It is a total indictment of the parents who clearly have no genuine concern for the long-term well-being of their children.

Being obese has cumulative problems, as it also means children and adults are not able to exercise easily, so the children do not walk to school, they play minimal sport, and instead they sit looking at computer screens or mobile phones as their 'exercise'. Some surveys indicate such screen watching by children can occupy more than six hours per day. This is not living. The pleasures of being fit, healthy, and agile are many, and we are stealing this from the young by cutting back on physical activities at school and failing to educate them into proper balanced and sensible eating habits. Selling school playing fields for building sites is not in the national interest. Many doctors have reported a considerable increase in young people with back problems caused by being hunched over electronic screens of phones or computers. Opticians similarly report a significant rise in young people with poor eyesight.

It also means that future adults will need more money to cope with obesity since, as for smoking, it is possible to estimate the cost to national healthcare. Obesity greatly increases conditions such as diabetes, heart diseases, strokes, cancer, mobility and joint problems, plus a very wide range of other medical conditions. In cash terms, this is £6 or £7 billion per year. In terms of the numbers of people, the annual estimates run at around half a million deaths from cancer and other obesity-related problems in the UK and in the USA. Further, their lack of mobility, and the strain put on all the families and carers, is immense but hard to quantify in monetary terms, but certainly it reflects a very selfish lack of consideration for these people, and an expectation that someone must look after them. Their failure to live sensibly destroys many more lives than just their own.

Numbers such as this (half a million related deaths per year) still seem to have little emotive impact, so a search for an alternative is to look at one side effect of Type 2 diabetes. Maybe we view a death as ending the problems, so they have gone away, and are no longer important. Perhaps a focus on obvious long-term factors, such as amputations, has more visible impact. It is a topic most try to ignore, so we should emphasize that in 2014, obesity-generated diabetes frequently results in major surgery and amputations. In the UK, these are running at 7,000 per year! For the young and mobile, one way they could relate to this number and appreciate the impact is to point out that it is *twice* the number of professional footballers in the UK. This is the current annual rate of amputations for the obese. Whilst these amputations are ongoing, they are not apparently newsworthy. Imagine the difference if the list included just one or two star footballers.

The numbers cited above are for 2014, but unfortunately the trend is downhill; current medical opinion is fairly consistent in saying that the number of people with Type 2 diabetes will rise towards 5 million within the next few years. Rephrasing this in numbers we recognize, this means roughly one person in ten of the UK population. Because the problem is regional and class-dependent (because of exercise and eating habits), the numbers may seem nearer one in six in many sections of the population! Supporting the cost of care and treatments on this scale is impossible. The problem must be addressed by a change in attitude, exercise, and diet. The root cause is a failure of many people to look after themselves.

With such a large population that is overweight, there is of course a considerable market opportunity to sell cures, fitness courses, or slimming pills that will lead back to the ideal perfect figure. As expected, these range from being excellent to extremely dubious, or even dangerous to health. Once again, technology can play a negative part in this. My particular example is that access to the Internet provides a vast range of tablets and slimming treatments. Not all advertisers are selling the products they claim, nor are they necessarily effective. TV programmes that have investigated such areas report that many of the slimming pills originate in countries where drug controls are minimal. Hence, not only may the products not contain the chemicals they claim to be using, but often the ones used instead are known to have side effects or are actively banned in the EC as being not just unsafe, but lethal.

The preceding examples reveal major problems that we, internationally, are allowing to develop, and it is easy to cite other areas such as excessive alcohol and drug use. The impact from their side effects is not always obvious as they are often hidden under an umbrella of other problems. Drunken behaviour, especially at weekends, is very obvious, but it is accepted as the norm rather than being seen as stupidity.

Looking at accident and emergency hospital admissions, the pattern of such weekend excesses is very clear. Perhaps a disincentive would be a 'healthy' fee to the hospitals for such emergency visits if the people are obviously drunk or drugged. A cost, such as their weekly wage, would certainly be a sobering proposal.

I am clearly of the opinion that instead of continuously pouring more money and resources into the health system, it would be far more effective, not only in terms of money, but also for the fitness, enjoyment, and pleasure of living, if there were a truly major effort to influence not just what we eat, but how we maintain our health. London was host to the 2012 Olympic Games, and the investments in sport, etc. were intended to inspire the nation to greater physical activity and fitness. By the end of 2015, surveys of over-16-year-olds suggested that the number of active participants in sport has *fallen* by half a million. This effort was therefore a failure, except for small numbers in about six minority sports.

I am being highly critical of the population in the UK, as I am using them as an example. The problem is actually typical of many advanced nations, but access to the costs and statistics are not always as easy. The USA certainly has parallel problems.

We need to generate a strong sense of personal responsibility for our actions, and to maintain our own condition, and enjoy and appreciate the rewards that it brings. I also think our political leaders have failed to grasp and understand the problem, so they take an easy option and keep promising ever more money and resources for a health service, in the hope of getting votes. The true problem is not that medical care needs to increase, but rather that people have lost the desire, responsibility, and ability to take care of themselves. If we regain that, then the need for hospital treatments and costs will plummet.

We also need to regenerate a sense of self-esteem and independence, because at the moment vast numbers of people assume, and expect, the state will fix all their problems, even when the problems are clearly caused by their own actions. An aggressive policy of publicity and education to help people change the way they live would therefore make immense economic sense. A fitter nation would gain far more pleasure from life; people could appreciate exercise and participate in sport (not just watch it). These are factors that would be truly valuable to society. A side effect would be a reduced need for medical services and social benefits. The only obvious losers include those selling junk foods, or unneeded pharmaceuticals, etc. I personally see this as an additional benefit.

The financial and health costs of our own failings and weaknesses

I am being critical of the existing system related to health, so it may be worth adding some costs, as assessed by Dame Carol Black in a recent government report, to indicate the scale of the problems and their cost to the UK nation. A thoughtful perusal of these numbers will need no more comment from me. The catalogue lists the following: 1 in 15 benefit claimants suffer from drug or alcohol dependency, with 2014 numbers as 280,000 for drugs and 170,000 for alcohol. These people have general health that is far below that of the rest of the nation. The related bills to the Health Service are £3.5 billion for alcohol, a further £11 billion in related crimes, and £7 billion in lost productivity. Obesity-related costs come out around £5 billion to health and £27 billion for the economy. I see this total (with tobacco) as being in excess of £65 billion.

Putting, say, 10 per cent of this number into education, training, and sport would surely make both economic and moral sense. My comment is appropriate in a book on the dark side of technology because medicine is one of the superb examples in technological advances and, because it is there, vast numbers of people assume it will cure our problems, rather than taking responsibility for their own destinies. We will never be 100 per cent successful, but we could do very much better.

Do we understand statistics?

Considering the thousands of drugs, medicines, and medical practices that have been tried, it is inevitable that some are genuinely disastrous, or for some the practitioners lack the necessary skills. Nevertheless, I think there are also a number of psychological factors that distort our appreciation of the services on offer. They are nothing to do with medicine, but are merely normal human mental responses. The first of these is that medicine resembles gambling, and we are willing to take chances, and have expectations of success, or winning, that are unrelated to the true statistics. Secondly, we are more worried by small potential losses and failures than small winnings or improvements. These are complex patterns of behaviour that have been studied in depth, and are the focus of the careers of many psychologists. Our behaviour is often irrational and easily skewed by the way data and information are presented to us. A fascinating book by the Nobel Prize–winner Daniel Kahneman (*Thinking, Fast and Slow*) certainly made me realize how often I jump to the wrong conclusions in ways that are absolutely standard behaviour. This difficulty is therefore very relevant in medicine, as there are few unequivocal medical responses where all people react in the same way. We differ, so treatments are more or less effective, or have side effects. Therefore various experts can appear to be totally opposed in the conclusions that they draw and we, the public, are confused. In many cases, the conclusions have been based on precisely the same information.

One fairly typical example relates to the use of hormone replacement therapy (HRT), which greatly improves the quality of life for a vast number of post-menopausal women. Since this is a treatment that worldwide has relevance for many million women, it attracts a lot of attention. In parallel with the benefits, one should certainly consider potential associated risks. Unfortunately, there have been a number of studies over the last decade or so that have drawn totally different conclusions. In some examples, as for the early studies of 2002 and 2007, the analyses are now seen to have been flawed, made with poor statistics, or are unrepresentative. For example, they may have included an unusually high percentage of those who are obese, or have been based on a limited social subgroup. Our psychology of reporting bad news, and remembering it, means these studies had high publicity.

By contrast, the 2007 study, from the very same organization, was not only positive in saying the incidence of additional cancers had been overestimated, but also showed that heart risks were less, and the debilitating condition of osteoporosis was generally prevented. This study had less media coverage and is mostly not remembered, as it was positive. A recent study again had high publicity as it said 0.1 per cent of users might develop a specific type of cancer. Such negative publicity always has media coverage, and it has caused nearly a million women to stop using HRT. The 99.9 per cent of women who benefit from HRT are ignored, or seen as unimportant.

One problem is that the 0.1 per cent is discussed as actual numbers of deaths (i.e. 1 death per 1,000 users), whereas those benefiting are overlooked. A proper comparison should be to say that for each unfortunate death, 990 women are leading far healthier and fuller lives. In general, the public, including scientists, respond emotively to a number of deaths, but fail to balance this if the good news is written as a percentage.

The puzzling question is therefore, Why should 0.1 per cent matter here for HRT when people ignore the 50 per cent death rate associated with smoking? We are illogical. Perhaps the weakness of anti-smoking campaign is that the health warnings say 50 per cent of smokers will die of smoking-related diseases. Many people are mathematically illiterate (or mentally unwilling to listen), so do not understand the relevance of 50 per cent. For smokers, we need to hammer home the message that 50 per cent means one out of two smokers dies as a consequence of smoking (i.e. use numbers).

The discussions of the value of HRT are certainly not conclusive and accepted by all practitioners. In addition to the negative potential of 0.1 per cent more cancer deaths of women using HRT, media reports give negligible coverage to positive evidence. A very specific benefit exists in the reduction of osteoporosis and that healthy bone structures are less likely to become cancerous than those with the disease. The results so far indicate that, over the ten years of the investigation, there appears to be an 18 per cent reduction in bone cancer deaths. This positive aspect of HRT totally swamps the negative ones. In hard numbers, not percentages, this says that for 1,000 women, one of them (0.1 per cent) will die from a related cancer, but 180 (18 per cent) will *not* develop bone cancer. The conclusion is remarkably clear, but ignored in media.

For HRT, the final, and, in many ways, far more serious consequence of the initial flawed studies is that doctors who were training at that time of the negative claims now have firmly entrenched views, and are unable to change them as a result of later positive data, and indeed will teach the ideas based on the original studies. This in turn will undermine a further generation of doctors.

Our failure to believe statistics can also be skewed by the way information is presented to us. There are endless TV programmes based in hospitals, which have large followings, and in addition to scripts basically about the interactions of the lead characters, there has to be window dressing of some medical action. A three-hour operation is clearly unsuitable (too long), as is MRI imaging (boring and too scientific), so there must be instant action items. One such is the need to respond extremely quickly if someone has heart failure. Manual attempts at resuscitation (called CPR, or cardiopulmonary resuscitation), or electric shock (defibrillation) are excellent for TV in this respect. They are instant and have visual TV impact. The local hero or heroine can save the life of a patient within a couple of minutes of the TV programme time, and everyone is happy. Both techniques, separately, or in conjunction, can indeed be successful, if (and only if) applied extremely rapidly after the heart stops beating.

The TV plots never offer the statistics, which broadly are that survival rates are around 6 per cent (i.e. out of 100 events a mere 6 patients survive, and 94 are dead) if the stoppage occurs outside of hospitals, and only slightly better when cardiac arrest occurs in hospital during other treatment, and medics are on hand. This 'success' rate needs a much harder and careful scrutiny because, if the blood flow to the brain is stopped for just a few minutes, then the 'survivors' frequently have severe brain damage, and may need permanent life-support systems. The latter option appears to be far worse than if the patient had died: it causes suffering for the family and patient, and there are very substantial ongoing healthcare costs.

Nevertheless, for most of the public, the image of CPR sounds like very good news. In my experience it is often taught to be used in situations where survival, without brain damage, is highly unlikely.

The dilemma of improving medical diagnosis

The dilemma of improving diagnosis is twofold, both for the medical profession and the public. New diagnostic techniques have advanced to the point where we cannot merely look at symptoms of illness, and attempt to treat them; we also detect potential problems. The majority of the public, and indeed a very large percentage of the medical profession, have not adjusted to this new situation. Instead, they are still stuck in the less enlightened philosophy that, if they find a medical problem or disease, then they must immediately interfere and apply all the maximum techniques, examination technologies, surgery, and drugs that might seem applicable. This is without taking the more crucial perspective that the treatments may not cure the condition, life expectancy may not increase, or quality of life may actually be made much worse. Additionally—and this certainly is not well appreciated—many diagnostic tests are inaccurate, not just in terms of failing to detect a disease, but far worse in that the results may offer many false positives for people, who either do not have the disease that is being tested for, or have it in a situation that it will never cause a problem.

The example of the prostate specific antigen (i.e. PSA) test to detect prostate cancer typifies some of the dilemmas and misunderstandings with a test that can genuinely offer diagnosis where treatment is needed, but equally is plagued with difficulties in raising investigations where it is not essential, or, even worse, where there is no cancer at all. The presence of a raised level of PSA is simple to detect, but at least two out of three men with a raised PSA value do not have any cancer. The excess may actually reflect athleticism or sexual activity. The consequent outcomes are, first, treatment can be successful, but the examination to distinguish between a cancerous condition and just a raised PSA level carries considerable risk. Incontinence, impotence, or both may occur in as many as 70 per cent of the men who are treated, but do not have the cancer!

Secondly, estimates suggest that in the elderly (say, over 75) it is extremely common, and if left untreated, because there are no major symptoms (merely the PSA reading), death will eventually occur from other causes. Very many men die *with* prostate cancer, not because of it. For doctors and patients, there is currently an excessive enthusiasm to operate based on the test, rather than actual symptoms. For those without serious symptoms, the ten-year life expectancy is no different from those with or without the condition. The real difficulty is that the testing and examinations can be extremely negative not only for those with such cancers, or those who will survive with the disease, but also for those who do not have it.

From the opposite viewpoint, the tests and examinations are essential and valuable in many cases. The negative statistics in this topic are that for those with symptoms, the treatments may help, but nevertheless prostate cancer can be the cause of death for nearly a third of them. Inevitably, the medical literature offers considerable conflict of opinion as to how to proceed for men with a raised PSA level.

A further problem with any medical diagnosis method is that the profession is naturally eager to not miss any condition, such a cancer. Therefore, the aim is to have tests that detect as high a percentage of afflicted patients as possible. However, with new technology and greater diagnostic skill, one can also sense potential *future* problems. Many of course will not materialize into any illness, but large sections of the medical profession believe we should intervene and respond to these latent conditions. The objective of skilled diagnosis is praiseworthy. But the level of consequent intervention is highly contentious, with factors ranging from political pressure, demonstration of state of the art (with patients merely as test subjects), high income from private treatments, to concerns that failure to intervene has the possibility of litigation if future problems arise. Matched with this is a public view that medicine is infallible, so people will demand unnecessary treatments.

Nevertheless, public and medical enthusiasm to detect diseases (e.g. breast cancer) with a high efficiency is invariably achieved by having an even higher rate of false positive results. This aspect of mass surveys is never clearly presented. Instead, mass screening is portrayed as harmless, accurate, and essential, and this is certainly not totally true of every type of screening.

This aspect of advances in medical technological has an incredibly dark side because it causes worry and anxiety, and destroys lives, relationships, and families even if no surgical interventions are attempted. The breast screening example is a well-documented example. There are several hundred types of cancer, and even for the case of breast cancer, many estimates say that perhaps 75 per cent of those detected are not life-threatening. For example, in the case of the elderly, cell division is slow, and the treatments used are often far worse than the disease. Additionally, X-ray screening or radio therapy are not passive techniques; they can and do induce cancers, etc. (current numbers indicate around 2 per cent are caused by medical X-ray treatments).

Far worse is the very high incidence of false positives, which can be as many as ten times greater than the number of breast cancers detected. For this vast number of people, there is anxiety, etc. and the possibility of surgical treatments. For these patients, there may be no consolation in eventually finding out that the problem was a misdiagnosis, or perhaps a benign cyst. There is also the indirect effect of people who become so worried by the possibility that they had a cancer, that the anxiety causes their health to collapse.

The preceding examples are well documented and discussed in extremely confrontational and obsessive terms by the members of the medical profession, to the confusion of the public.

A far newer candidate in this minefield of diagnosis is the recent ability to analyse in detail the human DNA. From it, one can make estimates of future health prospects, as numerous diseases and medical conditions have an increased incidence according to our genetic make-up. The possibility is not new. There are many examples of inherited diseases, or a propensity to develop them, which had already been recognized by looking at family medical histories. In extreme cases, people may wish to know if they carry undesirable genes which might be transmitted to their children, or maybe influence their decision to have children.

Less obvious is that within the last decade there have been rapid advances in recognizing faults on the DNA that can be linked to specific diseases, that may (or may not) develop later in life. For a modest sum, it is possible to have this prediction made by a very simple test. Most of those who have such analyses do so because they are already worried about their health and their future (e.g. are they likely to develop cancer or Alzheimer's, etc.). The fact that the DNA analysis is only a statistical indication of what the future may hold will be overridden by our natural instinct to look for the worst possible outcomes. The test results will always find some anomalies (every human is different), and this will play on our fears. For many, the very act of thinking about their future medical decay will shift them from being a statistical possibility to a self-fulfilling medical decline.

I have no idea how one can combat such attitudes, as they seem to be ingrained. Perhaps, for the majority of people, they should not try to peer into the future, but take the simple engineering guideline: 'If it isn't broken, do not try to fix it'.

Where next?

Medical knowledge, medicines, drugs, and techniques are expanding exponentially, and so are the associated industries and costs. The success stories are legion. A mere half-century ago, much of the current medical knowledge and practices would have seemed unbelievable or just science fiction. There is no way such progress will stop, but that does not mean there are not negative aspects. Nevertheless, I have focussed on examples of downside features driven by a general failure of people to look after themselves. The worst downside for me is that large sections of the population assume that no matter what they do to their bodies, all their difficulties and problems can be sorted out by modern medicine. The usual villains of smoking, obesity, drugs, alcohol, etc. blight the lives of millions and shorten the lives of many others. These types of medical problem need a total new assessment by government, with more effective education and actions that make people responsible for their actions.

Rather different difficulties are emerging from genetic developments that produce drug-resistant diseases, as merely by reducing antibiotic levels, etc. will not reduce the new strains of infection.

There is also the innate human urge to experiment and to genetically modify crops, people, and animals. In some cases, there has been progress, but as with all new and experimental ideas, it is impossible to know or predict what the associated downsides are. For example, anyone who found a genetic solution to a common cold would be hailed a hero, but if there is a mutation from the treatment, that means we lose some other attribute in later generations—we cannot reverse the clock and start again. If our attempts at improvement and cures mean we cause some premature deaths, then it is unfortunate, but if we endanger the entire human species, it will be disastrous.

In reality, the fantastic growth rate of medical knowledge means that individually we can only access and understand a minute fraction of the total literature and knowledge. This automatically means we will make increasingly more errors.

A non-standard view of the situation is to say that, if the database of medical knowledge is increasing faster than we can understand it as individuals, then effectively the technological advances mean we are increasingly more ignorant of the total knowledge base!

A final comment on databases and records related to medicine is that because people relocate and are treated by quite different medical experts at different sites, then there would be a real advantage in having records that are accessible from each site. Whilst the benefits are clear, the implementation of such a system is difficult, and so far, at least in the UK, the attempts have been a very large drain on the available money for the health service. The first trials in 2002 cost around £10 billion before being abandoned. Computational skills and expertise with large databases have improved since then, and a new proposal to make this transition to a paperless health service is budgeted at some £5 billion. In a smoothly running situation, this may be successful, but in times of crisis, with failures of communication networks, then it will generate even more problems. Security and confidentiality may similarly be compromised.

9

Knowledge Loss from Changing Language

Language and why are humans so successful

We are not the only intelligent animals, so what is the key factor that has allowed humans to be productive and innovative? For me, the crucial distinction between us and other creatures is our ability to use a sophisticated language. Without it, we might have remained a successful pack animal, such as monkeys or wolves. In many ways we are still pack animals. We still need support from the rest of the group to hunt, to raise young, and to pass on skills by example. We have spread around the world into different climate zones, but so have many other creatures. Some, such as the wolf packs in Yellowstone, actually thrive in the harsh winter conditions.

Other animals can also use tools, with monkeys and crows as very different but classic examples. They also have the ability to plan a sequence of events to provide later gratification in terms of food. Their skill limits are set by their structure (e.g. beaks are not as convenient as hands) rather than intelligence. Other primates have good manual skills, but somehow we humans have achieved much more.

Intelligence, and communication by sound, is not rare; many animals, both on land or in the sea, convey quite precise information. For example, meerkats have calls that say which types of predators are threatening them. Parallel examples exist for marine creatures such as dolphins, whales, and octopi. Despite this, in terms of language, our level of complexity and sophistication is unique.

A further advantage for us is that we benefit from a lengthy immature stage where the young are dependent on the older members of the tribe. These formative years have nurtured the use of language as well as learning practical skills from several generations. Language, not brain size or intelligence, has defined our break away from the limitations of other species. This is not speculation, as during the last century explorers have found isolated tribal societies that appear to us to be effectively still at the level of the Stone Age, but despite their lack of technology, they have detailed structured languages.

Language and a relatively long lifetime have allowed experimentation to manufacture tools that helped us in hunting and survival. Since we lack strength, claws, or jaws to catch and kill our prey, the production of flint-tipped arrows and knives was front-line technology that had a dramatic impact on long-range hunting in safety. This must have felt like real progress for those early people, and it was 100 per cent positive. Reality was of course very different when viewed over a wider picture, as successful hunting progressed to sow the seeds to hunt larger animals. For the animals, this was a disaster, as in many cases it was the downward path to their extinction. Some, such as mammoths, may already have been struggling with a changing climate, but others, particularly large species with a long life and low reproduction rate, were rapidly doomed by our hunger, greed, and sense of survival.

The final step in our success story has been not just to maintain knowledge by speech, but to devise ways to write and transmit our thoughts from one generation to another across the globe.

Decay of language and understanding

Ability to record, store, and transmit information, ideas, and images makes us unique; without it we would definitely have never reached our present level of knowledge and technology. Access to existing knowledge, as well as past experience, skills, and information, are therefore very high on our list of priorities. Nevertheless, the range of facts that we would like to discover and use is very wide, and there is no unique pattern or recording system that can do this for us. Equally relevant is that both language and storage techniques are not permanent. Particularly with modern technologies, the storage aspects may survive less than a generation.

Often the window of opportunity to access specific facts and knowledge is very limited. A typical situation, which is highly personal and immediate, is to have information about our own family history and past situations. When we are young, we never really care about such history, as there are normally relations and friends who may have knowledge about it, and, at that stage in life, we are too eager to experience the present, and plan or explore the future, than to worry about the past. This is a very standard mistake as all the family photo albums, the anecdotes about our parents, grandparents, or second cousins would be accessible to us if we only stopped to question the relations who are still alive and alert enough to fill in the details. Once they are gone there is no way we can gain from their experience, nor add names and dates to the photo albums. It may even be that there were interesting family skeletons in the past that they would have been willing to disclose, had we asked.

The moral of this example applies more widely than just to family records. It is saying we must gain (and transmit) as much knowledge as possible, whilst it is still accessible, especially where living sources are involved.

My parallel advice for older generations is to deliberately volunteer the facts to the younger ones, whilst we can. Not only does this mean our ancestors will be remembered, but so will we. Humans have a vain and egocentric streak and we do not want to be forgotten.

There are two main reasons knowledge and information from earlier generations have vanished. The first is that we may have records, but they are in languages that either no longer exist, or in languages that have now evolved to such an extent that we can read the words, but fail to see them in the context they were written. The second cause of information decay is that it was produced on material (e.g. parchment or paper) that has not survived.

The latter problem is less obvious, as we are being conditioned by all the media and marketing that our knowledge base is expanding and it is becoming ever easier to access it from the Internet. Further, we can track family histories and have instant CCD (charge coupled device) camera and mobile phone photos that we put on our computer, CD, or hard drives. So we assume this should guarantee that the next generations will remember us. Unfortunately, this is totally false. The reality is that the very same rapid advances in computer technology are making current computers, software, and data storage systems obsolete. They are continuously replaced by newer versions. Whereas we may have some faded photos of Victorian relations we, in the early twenty-first century, may be lucky if in 20 years' time we ourselves will be able to retrieve our own photos in a format that can be electronically read.

The technological advances that I will describe are revealing an unequivocal and easily demonstrated pattern that progress is bought at the expense of losing historic data and information. Indeed, our pictures may be inaccessible long before our memory has failed. Similarly, Internet information depends on it being maintained and operated without extortionate charging, a feature that is likely to occur in order to limit the traffic once the technologies can no longer cope with the ever-increasing demands.

Information survival

Broadly there are various types of information that (a) we absolutely need, (b) would definitely like to access, plus (c) material that is interesting, but we could survive without it. I will give examples of each type.

In the prime position of category A are data related to our bank accounts, birth certificates, passports, driving licences, taxes, insurance policies, postal and email addresses, and phone numbers of friends and contacts. Without access to them we would rapidly head into a chaotic state and might not even be able to buy food and other goods. Other data, such as our medical records, are hopefully needed less frequently, but their loss could raise problems. For all these types of information, we need reliable and accessible data that is not easily destroyed or corrupted by other people for criminal or accidental reasons. A mere 50 years ago, we accepted responsibility for such data storage. We kept paper records, which were secure from electronic spying, but nevertheless were vulnerable to fire and theft. Technological progress from paper to electronics offers many benefits, but an increasing uncertainty in terms of security, or, in some locations, poor electronic availability.

Perhaps in this first category of key items of knowledge I should include the fact that many skills of craftsmanship and manufacture are not, and cannot, be written down. These are the personal skills that are learnt via training and apprenticeships, and they need verbal contact. They range from metal and construction techniques to playing musical instruments. As we move from actually learning manual skills, to just sitting at a computer, we are destroying irreplaceable knowledge. The change can be divisive as we automatically feel superior to those who do not have our own skills, but the craftsmen, from whom we could benefit, may be less computer literate than the young. Therefore we may fail to listen, which in turn may cause a rift between the generations. It is not just the young who will undervalue skills with which they are not familiar—it is a two-way problem. New technologies are thus divisive in the key step of acquiring many types of skill, data, and verbal traditions.

Category B refers to data and information that we would like to access. This is characterized by material that once might have been available in books, journals, libraries, or manufacturers' catalogues. Similarly, for many items it was possible to visit a shop and have a direct discussion with staff that understood their product, and could give helpful advice. With Internet purchases, this personal touch is lost, and we can never tell if one manufacturer makes a better product than another, as viewed from our specific needs and preferences. The electronic route may appear to be cheaper, but if there are many alternatives, decisions are harder. We also lack sight of the items until they arrive, so cannot judge a true colour, texture, or quality of items such as clothing or furniture. Further, many products cease to be produced. Many skills, data, and items vanish—they are lost and cannot be replaced. This is irritating and unfortunate, but not immediately life-threatening.

Category C is more in the realm of history where we want information and ideas that were discovered and discussed in earlier times. Some such data may link to skills that are now obsolete, but we wish to duplicate; for example, in a restoration project. Other items may be purely for historical interest. Finding the source material may be difficult, and the historical records may not even be in a language we understand, so we will rely on translations or other commentators. Realistically, if such information is lost, then it is sad, as it relates to our cultural heritage and understanding of how civilizations have developed.

Whilst I personally find many historical events, documents, painting, and music are interesting, I am also very unhappy that we fail to have learnt much from past history about the negative side of human nature. At all times, we have continuously waged wars, enslaved people, or persecuted them in the name of progress, territorial greed, or religious fervour. Since this is a message that we really need to learn, then I feel it is equally essential to have the historical records available to us as the immediate items in my prime category. If we learn, then perhaps we may aim for a more considerate future. (Clearly I am an unworldly idealist.)

These three categories of information obviously decay for a variety of reasons, and I will discuss the problem of language loss in this chapter, and, in the next, look at the survival and decay of the materials on which we have recorded information.

Lost languages

I will start by considering information that was hidden in languages that may no longer exist, as this is a key problem with historic documents (or stone carvings, etc.). Without it, we may have the text, but be totally ignorant of the content. Only when we can translate it do we realize it says 'Buy extra socks', or 'Here is the location of buried treasure'. The degree of language loss can be varied; so, for example, ancient Greek, Latin, or Old English may have similarities to modern variants, but in other cases entire civilizations have vanished without any direct link or translation available (e.g. as for the Minoan civilization that was wiped out by the volcanic explosion of Santorini).

Languages, like Latin of two millennia ago, have become dead and unchanged except for specialist purposes such as religion, law, botany, and medicine. The survival of written Latin may additionally have been prolonged beyond its use-by date as it was a required entry qualification into some universities. By contrast, spoken Latin evolved into Italian and other Romance languages such as Spanish or Romanian. The value of Latin as a unifying European language initially appeared as a military and administrative factor in running the Roman Empire. Such a pattern has been replicated many times with other languages, where military conquest has spread over large regions, and the winners have imposed their language for administrative purposes.

In many ways, the conquered lands may have benefited, as large regions could then communicate and eventually break into independence. Two familiar examples are for the Indian subcontinent and the Congo. In both cases, the lands were so extensive that they had each supported several hundred local languages, and none were acceptable as the administrative dominant winner, because it would have too strongly favoured one region or tribe. The imposed languages therefore aided unification into new countries, but it has resulted in far fewer of the smaller language groups surviving. Indeed, often administration has been simpler when the regional languages were deliberately suppressed.

Current estimates are that around 25 per cent of existing minority languages across the world will be extinct within one generation, basically because of globally driven technologies. This 25 per cent number is despite the fact that definition of an active language is absolutely minimal. Language survival only requires that it exists, and is used, within just *one* community. This is certainly far less than I imagined would be used in such a definition.

A more recent instance than the Latin of the Roman Empire is the case of France at the time of Napoleon, The country, although nominally called France, was supporting at least 40 languages and dialects, each of which was virtually unintelligible to the rest of the country. Unification came via technologies such as a detailed and accurate cartographic effort, with maps in a single language and spelling (the Parisian version), a rationalization of weights and measures instead of different values in every district, and an imposed single language for the administration. Definitely these were advances from many aspects as they enabled political, economic, and career opportunities, and produced a nation with one identity. Nevertheless, they spelt death for most of the local minor languages. Unification similarly undermined many of the surviving dialects because new words, which were introduced for administrative or technological reasons, entered all the dialects equally.

Regional variants and dialects still persist, but they are steadily being homogenized via radio and TV.

Language and technology

Returning to the nineteenth century, another factor for globalization driven by technology came from the railways, both internally and internationally. They had an impact on both dialects and import of words from one language to another, as suddenly it became a simple matter to travel long distances, and with it to hear and inject new words. The hoteliers and shopkeepers saw big profits in being able to communicate with rich tourists, and did not realize how they were unifying and globalizing language.

New technological, or imported, word corruption is a global pattern, particularly for tribal languages that are spoken but not written. It contributes to their dying at a high rate as the populations of native speakers decline. Total loss is often slowed as, for sentimental or regional reasons, small groups try to preserve their ancestral languages. In the UK, Welsh is happily surviving (albeit with modern technological additions); there are radio and TV channels that will support it. This is excellent as it may guarantee some stability, but it is unlikely to spread elsewhere to non-Welsh people. Exceptions can exist, as exemplified by an isolated Welsh mining colony in Argentina that still uses Welsh.

In the USA, many of the languages of the Native American Indians have declined to this local category, and most have been significantly corrupted by the import of modern words. This is unavoidable as a language such as Sioux would not intrinsically have had any words for modern technology (electronics, computers, cars), and, once corrupted, loss sets in for the number of people willing to speak only the original versions.

By the twentieth century, broadcasting, TV, and films had an equally dramatic effect. I know many older people who remember that the British dialects were so strong that it was impossible to hold a conversation between people from different regions, even if they thought they were speaking the same language. Professor Higgins's comments in *My Fair Lady* (*Pygmalion*) were very apt at the time of the original Shaw play, which predated radio. Regional accents still exist, but they are far weaker than previously.

The internal divisiveness of dialect and language has certainly not vanished, and the splits come not only by region but by education, social class, and activities. Equally they are obvious in the case of invasions and colonizations. In the Norman invasion of Britain in 1066, the new ruling class spoke French and the losers used Anglo-Saxon. This therefore linked language and social divisions with long-term consequences that are still apparent. In broad terms, the modern affluent classes use the longer words derived from Latin and French origins, whereas the rest have far more short words based on Anglo-Saxon or Germanic routes. The positive view for the UK is that our history has given us a rich and diverse vocabulary, but we mostly use only a section of it. Our roots are not easily disguised as the class split is often deeply ingrained, and although people may have attended the same university, and speak with the same accent, their origins can often be revealed by their choice of words.

Other technologies, such as international flying, need a universal language, and for historic reasons this is English. For a major language such as Mandarin (or Cantonese), the impact of mobile phones has resulted in the need for simplification in writing text messages, and the solution is Pinyin, in which the characters and sounds are expressed in Western text. Hence typing a few Pinyin letters offers a predictive set of Chinese characters. Overall, it offers a much faster route than attempting to construct complex characters on a mobile. A similar approach is now used in some Chinese schools, but unfortunately the same set of

Western letters is not identically matched to the Western sounds. However, other benefits from the use of Pinyin and predictive text technology are that writing (via a keyboard) is simple and of real benefit to those with unsteady handwriting.

The technology of typing had a major impact on the teaching of writing, and the majority of people now lack the expertise that was once the norm. As one who relies on the keyboard, I am amazed, and jealous, of the superb examples of legible 'copper plate' handwriting of the nineteenth century. For them, it was equally necessary to have such legibility in other areas such as account books. The ledger entries have provided a record of daily life over the centuries for everything from commerce to daily housekeeping. Historically, this is highly informative and a feature that will have vanished for future historians and archivists who will have no ready access to such computer-based accounting. The pre-decimal arithmetic in the ledgers is equally astounding to modern generations who rely on calculators, even for simple addition. I doubt that many modern shop assistants (or managers) could cope with mentally working out the cost of 17 items at three pounds seven shillings and five pence three farthings plus . . .

It is inevitable that languages evolve, and become enhanced or altered by contact with other languages, and this sets a finite lifetime for them. As ever, technology plays a role, whether by enabling warfare and conquest, or exposure to other languages via writing, books, movies, TV, radio, travel, and global Internet communication. Some countries, such as France, make a great effort to preserve their tongue, whereas other nations are happy to be multilingual, incorporate foreign words, or both. The rate of language loss will partly depend on the initial numbers who use it and the importance of the society that uses it. The clear pattern overall is that there is a reduction of the number of mainstream languages.

Language evolution

Even the survivors are of course never static, but equally alive and mutating on a daily basis. There is no clear pattern of what defines a 'half-life' during which most of a written version of the language can still be generally understood. From a parochial viewpoint of citing only changes in English in Britain, then it is clear that pre-electronics (so, say, 50 years ago), written documents by a Victorian author, Charles

Dickens, would generally have been intelligible to, say, 80 per cent of the population, except for words that had vanished from daily use. Moving back 500 years to the plays of Shakespeare would cause more language problems, but still some 50 per cent of the text would have been recognizable to most people. However, stepping back a little further to the writings of Chaucer would be far more problematic. Here I doubt that 25 per cent of the general public would truly understand (even if they have the broad spirit of the subject matter). So this would put an evolutionary half-life for English around, say, 500 years.

To emphasize that loss of understanding is just as obvious on very short timescales, I will give two examples. The first is that spellings, as in Dutch or German, have changed substantially within a lifetime; handwriting before the Second World War used Sütterlinschrift, which is mostly unintelligible to a modern generation. The second example is from even more recent times. Certainly in the 1970s and 1980s, any good secretary or reporter would be totally skilled with shorthand. However, even if the notebooks have survived that were reporting key events, virtually no one can now read what was being said.

In the modern UK over the last century, there have been large influxes of immigrants from across the world, and for them the earlier texts will be effectively a different foreign language, so a modern survey would estimate that the earlier English has a much shorter half-life than my first guess at 500 years, and many sections of the population would struggle with English that was written, say, 80 years ago. Even shorter timescales apply to dialect and teenage slang. Familiar words may be used, but teenagers attribute totally new meanings to them. The survival timescales of fashionable teenage variants have a half-life equivalent to the time spent as a teenager. Future historians will be bemused that what was 'hot' became 'cool', as well as overuse of multipurpose words such as 'like' and profanities.

Overall there is a worldwide trend to fewer and fewer languages, and although almost 10,000 languages are still in use to some extent, half of the world's population use only ten of them. At the other end of the scale, around 5 per cent of the population is spread across about 7,000 languages, often with fewer than 1,000 speakers. At this level loss is inevitable within a few generations, not least because overall there is much more mobility away from one's original village (or country), especially by the young. In fact, I am citing optimistic estimates, as many linguists predict that between 50 to 90 per cent of minority languages will be extinct by 2050. This modern loss of languages is frequently driven by changing technologies of transport and electronic communications.

Reading and understanding past languages

The first challenge is that the written material from a former era may neither be in a nice simple modern writing style, nor in a language that is still in use. We currently (in English) are using just a few letters to form words. Mostly, the spelling offers a reasonable guess at the sound that is spoken. Some modern languages, such as Spanish, are excellent in this respect, but both UK and American variants of English are far less reliable in moving from written to spoken sounds. Troublesome letter groups, such as 'ough', are a particular hazard, so a sentence such as 'The tree snake from Slough fell off the bough because his skin was rough and he coughed when trying to thoroughly slough it off' is a speaking minefield for a visitor. Equally problematic is that the same words are used in many countries where the pronunciation can be very different.

So if we have problems in a current language, we must expect far more when dealing with written scripts where we may never have heard the true sounds or the cadence of the speech. This is a serious weakness, as the voice conveys considerable information. Even in modern languages some are tonal, for example, the same word in Chinese can take four meanings as they use four standard tone patterns. So the written form of the same word, four times, can be spoken to make a sentence of four totally different words (hence a phrase such as 'father pulls eight targets' would be spoken with the same little written word, but with four tonalities).

In terms of writing, the earlier Western scripts were extremely varied and have ranged from incised lines in clay or rock to Egyptian hieroglyphics. Pictogram origins are equally obvious in the Far Eastern writing in, say, China or Japan. Therefore, the first step is to try to convert the technology of the written symbols into a modern format (and hope there are not hidden tonal factors).

To a large extent this is a decryption problem that appeals to code breakers, and some modern attempts are now being made via computer software. But normally the best approach is to search for parallels where the same text is written in several languages, ideally where one of them is moderately recognizable. For major empires, such as the Egyptian, Greek, or Roman examples, they had clearly formulated laws and official documents that were translated into the language of their regions and colonies, so often there is the chance of parallel translations. One of the most cited examples, which helped solve the Egyptian hieroglyphics, was the basalt stone tablet found at Rosetta in 1799. Inscribed on it was a document from Ptolemy the Fifth in about 196 BC that was repeated in Greek, Demotic, and hieroglyphics. A clear winner!

Similarly, ancient pictograms were found in the Middle East that evolved into the script called cuneiform. The clay tablets used have survived more than 5,000 years and the code was broken initially by a German school teacher, Georg Grotefend, who found an inscription relating to the kings of Persia. Further progress was made by a trilingual inscription written in languages of early Persian, Elamite, and Akkadian. The last two may not even be languages most of us have ever heard of. Finding and breaking such language codes with simultaneous translations is very fortunate and it has led to success with related languages using similar scripts. Before being too excited with the attempts at translation, we must not forget that we have only found the words. This is still a long step from adding implications and nuances that would have been read into them by the original readers. In our own experience, the same text from different friends, or different politicians, may be interpreted very differently.

Even without the multilingual help of ancient government decrees and the like, other scripts, such as Linear A and Linear B, have been understood, and just needed a persistent and inspired translator. Linear A was a mixture of syllabic and ideographic writing found on Minoan Bronze Age pottery in Crete and dated from around the fifteenth century BC. Linear B was on clay tablets and was written somewhat later, and is now assumed to be the early version of Mycenaean Greek. All these examples show written material was available at least 4,000 years ago. Literacy from other ancient cultures, as in the Far East, has now become of wider interest, even to Western society, but has been hindered by greater language barriers. They can offer truly historic insights, as there are Chinese inscriptions on jade and clay dating from nearly 3000 BC.

If we realize that the total world population over 4,000 years ago was minute compared with today (~7 billion at present), and realistically only a small percentage of them were both literate and living in areas where their written material might survive to the present day, then this means that cuneiform and hieroglyphs were effectively being written by a very tiny 'relevant literate worldwide' population. It is therefore amazing that such a large amount of material is available to us. The artefacts have survived in part due to the ruggedness of the stone and baked clay. Estimates of world demography are difficult for the ancient world, but in 3000 BC the total world population was probably no more than 20 to 50 million (i.e. just a few times the size of some current conurbations). As a sensible guess, only a few per cent of the population would have been literate and maybe only 25 per cent were in areas where it has been possible to make any systematic archaeology. So at best we are searching for writings from populations of not more than 50,000 to 100,000 people in a particular generation! The success rate is therefore surprisingly good, and far in excess of the survival rate that we could contemplate from modern writings of several billion literate people. Indeed, long-term data and literature storage is probably undesirable for much of our efforts, especially the trivialities we express in everyday communications via tweets and blogs, etc.

The challenge of translation

Finding a modern equivalent to an ancient word is still some distance from understanding the language and true meaning. Often translations have emerged from political or official documents, but we only need to listen to our current politicians and lawmakers to realize that the words have many interpretations and that the reality of meaning and action are not 100 per cent related to a text. Indeed, from skilled politicians, ambiguity is often intentional.

Translation skills are variable; in many cases the translations were commissioned for a particular purpose, so the objective will have skewed the meaning of the original, or the original may be in a dialect or style that is not appreciated by the translator. In documents such as those related to religion, there have always been heated ongoing debates about accuracy, bias, and deliberate distortion of the different translations from the original languages. Often several layers of translation are involved, and it then becomes very hard to persuade anyone that the text they are familiar with is in error compared with the original. I recall once hearing a professor of Aramaic saying that the earliest Greek translations mistook a word for 'artisan' as 'carpenter', rather than the very similar Aramaic writing with the meaning 'mason'. He may well be right, but I am sure few modern Christians will believe it, as it is too deeply ingrained in their traditions over two millennia. We should also recognize that many such religious works were deliberately kept in languages used by a very select group, precisely (as publicly stated explicitly in Tudor times) so that the masses could not read them and think for themselves.

Religious texts offer many examples of intentional suppression or change by translators or those people who control opinion. In the case of the New Testament, there are now officially four gospels. These were selected by the Bishop of Lyon (Irenaeus, who lived from 140 to 200 AD), as they have a moderately consistent presentation. It was a very pragmatic decision as the four are similar, whereas many of the others offer quite different perspectives. Irenaeus lived at a time of intense Christian persecution, so any material that raised doubt for his followers, or was a weakness to be exploited by the Roman oppressors, was best ignored. This view was reiterated in 1546 by the Council of Trent. However, religious scholars are aware of many other writings, perhaps 30, that existed that had been proscribed, or trashed, because they were in conflict with the choice of Irenaeus. In many cases, they offer quite varied interpretations and opinions. There are many texts, but true knowledge is lost or clouded, and finding an absolute truth is impossible, not only from what is written, but also in the way we interpret what we read.

Translation 'errors' that are politically distorted to place one group in a more favourable light or to offer legitimacy to an invader are inevitable. The same events may be written as an 'invasion' from one side and presented as 'liberation' by the other. If only one view is found and translated, then we can be seriously misled.

To a historian, ancient documents are fascinating and, certainly if in Latin or Greek, are relatively readable in terms of text, and broadly so in terms of meaning. However, language is not passive, but it changes with time, dialect, and region, and is strongly sensitive to the culture and background of the people involved. The same is true for modern languages, and, even for classic English, I have already mentioned that for literature, such as Shakespeare or Dickens, there are real problems for most people in fully understanding the words. Going a little further back to Chaucer, the general public would have severe problems understanding the text or speech. Trivial examples are words that have totally vanished from current usage, but more problematic are words that have changed meaning and nuance. For example, the Elizabethan 'presently' meant immediately (i.e. at once, in the very present time), whereas by the twenty-first century 'presently' means sometime in the fairly near future.

Language evolution will even be obvious if you attempt to do a crossword in your favourite newspaper, and compare the difficulty in solving a modern puzzle with one from 30 years ago. The older one will have clues with nuances and context that make this very difficult.

Examples of differences between the variants of English, Australian, or American, or Spanish and Mexican, etc. can be equally misleading both in meaning and culture. This is especially true when apparently the same word exists. For example, a 'discussion' in English is a friendly consideration of some facts, whereas in Spanish it is an argument, but in German an 'argument' is a rational discussion. Similarly, 'quite good' in English (UK) is very positive, but in the USA it is only lukewarm, or critically negative. This is a subtle style problem that is usually overlooked in films and TV, but it alters our perception of the plot and the characters.

From my own experiences, I had a number of problems in the USA when I used, or heard, colloquial phrases. At least from my side, I could actively suppress their use, but I am sure I have often misunderstood informal American chat.

Other ongoing problems of different meanings for the same words can be remarkably important. In our use of numbers, we happily use ten, hundred, thousand, and million (10, 100, 1,000, and 1,000,000), and in scientific articles we use a shorthand where we indicate how many zeroes there are via 10, 10^2, 10^3, 10^6. For the next big number, the UK used to use a billion which was a million million (10^{12}), whereas in the USA, one billion is only a thousand million (10^9). Somehow in the recent past, the UK has taken over the US definition of a billion. It is not clear how many people know that we changed, but it is important, as the numbers differ by a factor of a thousand! Very recently I was shocked during a talk at a scientific meeting, where people were using words, not the written 10^9 or 10^{12} notation. I suddenly realized that in a multinational discussion, the Spanish and Germans were using billion to mean a million million (10^{12}), whereas others were implying a thousand million (10^9)! Politicians never use the scientific notation, and I wonder how many errors are being made because of this.

Such problems are hidden when we have little familiarity with the magnitude of the numbers involved. I have mentioned the difficulties caused by the use of two temperature units (Celsius and Fahrenheit). These scales set a divide between Europe and the USA, and within the UK between old and young. Elderly people will instinctively know whether 50°F or 70°F implies the day will be cool or hot, but they are far less secure with numbers of 15°C and 25°C.

Language and context

Original context is impossible to reproduce, and modern rewrites of old plays and books for film or TV dramatizations change the attitudes of the characters so that actors speak, act, and respond with the social attitudes and behaviour of the twenty-first century. Effectively, the attempts at rewrites to modernize the text are little different from a translation in which the original language changes are obvious, but the background culture and attitudes are blurred or lost.

Any source material with emotive content that involves, say, discussion of class or political status, religion, different races, or the role of women in society, will be extremely difficult to correctly translate, as these involve social rather than merely literal factors. So, translation is feasible but real in-depth meaning can be quite difficult, or even wrong. The 'decay' of this deeper level of information with time is inevitable, and often is less than an average life span.

Even in music of operas, etc. the prevailing attitudes of the audiences would have interpreted the libretti in ways that now seem totally unbelievable from our viewpoint. Classic examples in the poor attitude towards women are seen in famous operas such as *La Traviata*, in which the heroine is forced to sacrifice her happiness to satisfy the respectability of her lover's family. The text may well have truly reflected the attitudes of the time. Views on morality differed greatly between the permissible behaviour of men and women characters, which now seem very unjust. Humour is inevitably a problem, as puns and jokes, as in Shakespeare or Gilbert and Sullivan, were highly topical, and are totally lost on us. Gilbert and Sullivan plots, settings, and staging, as well as the words, were making very strong and pointed social comments that we have little hope of understanding 150 years later. Society advances in part from technological improvements, so rewriting for TV, books, and films means we are irreversibly and steadily clouding our understanding of the past and forcing it into our current view of morality and lifestyle.

Information loss in art images and pictures

The logical assumption is that if information is fading away from us in language, writing, and literature, then the same pattern is likely to occur in other areas. Paintings and carvings predate writing; the oldest known cave paintings are from some 30,000 years ago. Their survival has been possible because of climatic conditions in caves, as well as the lack of sunlight, which could fade the colour of the paints that they used. The paintings are often simple depictions of animals, so tell us a little about the animals alive at the time, but we can only guess if they were paintings purely for pleasure, if they had religious significance, or if they were related to passing on hunting skills. Other art forms of carving and painting on rock, wood, vellum, paper, canvas, etc. have all the limitations and sources of loss as for written material. In fact, in some ways they are worse in terms of survival, as painting is dependent on colour, and the dyes and pigments that were used were rarely stable over long periods of time. Additionally, canvas and wooden frames are prone to attack by insects, and large paintings in churches or castles were only maintained when the site was wealthy and fashionable.

With more famous pictures, there is the equivalent of changes introduced by translators in literature, but here the changes are from repairs and restoration, or those who felt improvements or alterations were needed. Particularly with nude figures, there have been times of religious concern, intolerance, or prudishness that had clothes added to the original paintings, or parts of nude statues removed. I see this more as data loss rather than information loss.

Renovations and repairs are costly, so are most likely to have occurred with works by famous artists. A truly classic example of poor restoration attempts is seen in the painting of 'The Last Supper' by Leonardo da Vinci. He made the original in 1494–98 with a technique known as tempera on gesso that was glued on to a north-facing wall with mastic. Unfortunately, from the viewpoint of preservation, it was in a dining room next to some kitchens. The advantage of tempera on gesso is that it offers finer detail and colour than the more common fresco wall paintings. The tempera on gesso route can be painted with more care and slower than a fresco on damp plaster. Unfortunately, the painting surface did not bond well to the damp wall, and it was made worse by proximity to the kitchens, and almost immediately the picture started to have problems. Subsequent use of the room as an army barracks did not help, and by around 1642 the picture had virtually vanished. Nevertheless, copies had been made, and since then there have been numerous attempts at restoration. Some were highly imaginative. 'Restorers' used different types of paint, and modern analyses show they changed the colours, redirected the gaze of different figures, and overall it is now only possible to guess at the real impact of the original.

By contrast, just a few years later, in 1500, Raphael painted a magnificent scene called 'The School of Athens'. It is still in excellent condition, but although we can see the craftsmanship, we lack the understanding of the content. The figures were supposedly images of current or previous artists or philosophers, but there is no legend beneath the picture to say who is which one. Art historians agree on a few of the depicted characters, but many are lost to us; so we can only see the work as a painting and have lost the social comment that was intended. This is typical of much of the art one finds in museums, as the paintings will contain references to mythology, local characters, famous battles, or other events that would have been recognized when created, but now our generation is ignorant of this background. Types of flowers indicated purity or fidelity, or a red cloth might indicate a brothel. All this encoded information is lost, even if we recognize the skill of the artist. The speed at which such understanding fades away is increasing as modern societies have a more technological content and education, and less interest in ancient mythology and ancient historical events.

Music and technology

Language and cultural evolution are apparent not just in literature and painting, but also in music. Initially, music was totally linked to language and culture, as the only way it could pass from one area to another, or from one generation to the next, was by personal contact, hearing, and memory. This situation existed from the earliest times, and there are pictures that include musical instruments from Assyrian, Egyptian, early Chinese, and other cultures. There are similarities in the instruments portrayed, with lute styles suitable for use with singing in a small group, and big drums or brass instruments for use in battlefields. The music they played is of course lost, as are the songs, ballads, folklore, and their words.

Technologies have driven many advances and changes in the dissemination and style of music. A very early example was the introduction of written musical notation, which meant that church music could be sent from one place to another. So if an elderly monk, who knew a particular piece of church music, died, then the music was not forgotten. Credit for such innovations in writing music includes input from a monk, Guido d'Arezzo (995–1050), who wrote a notation that speeded up learning, and could be carried between monasteries. Equally important for Western music is that d'Arezzo introduced a semi-standard musical scale that has defined much of our modern musical language, and gave us the sol-fa naming of notes in a scale.

Musical scales are just as varied as spoken languages. Those with any interest in music will probably know that Western music uses an octave with 12 semitones (e.g. an octave includes all the keys from, say, C to C on a piano keyboard), whereas many other cultures use a five-note, pentatonic scale (roughly spaced as just the black notes on the piano). In broad terms, these may partially reflect tonal sounds of language between the west and east. However, in detail, there are dialects. For example, the Scottish bagpipes use a unique 'dialect', 12-semitone scale variant of the piano keyboard. Jazz tuning is also different. There was (and is) a serious problem that singers and string players instinctively adjust the tuning very slightly as they play in different keys, and notes such as C sharp and D flat are different.

One can say a tune is written in a particular key (for example, the key of C). This defines the frequency spacing of the notes that are sung in that key. If the singer changes key (for example, because the original was too low or too high for their voice), then they use a totally different set of notes. The voice is flexible, so it is not a problem. However, on a keyboard instrument (such as a piano), there is only a single note to serve all the variants demanded by the various key signatures of the singers. This was a serious problem with unpleasant sounds, but technology, in the form of mathematics, was used to resolve it and define each semitone in equal frequency ratios. An octave ratio is 2:1, so for 12 semitones each ratio has to be the twelfth root of 2 (~1.059). It was progress, mathematically elegant, but with *every* note being out of tune! (This is technological progress.) In fact, this is still not enough distortion, and piano tuning is not uniform across the entire instrument. We have adjusted to the compromises and associate them to the characteristic sound of a piano.

Like languages, many scale tunings exist; websites describe, and offer the sounds, of a total of some 50 different scales that are in use around the world. For those who are poor singers or players, this is encouraging, as they could claim they are just experimenting with one of the less familiar scales. More realistically, there is no absolute musical scale. The only truly common feature is that the octave has the higher note at twice the frequency of the lower one.

Technology has introduced a vast range of improvements and innovations in musical instruments, as well as stimulating new ones, and via electronics of broadcasting and recordings allowed us to hear music of foreign regions and made the music of skilled performers accessible across the world. Familiar examples of change and improvements are that pianos developed with more power and a wider range of notes, so they could be played in larger concert halls than the early variants of the harpsichord. Trumpets moved from simple hunting horn–type notes to have keys, and then valves, so they could play chromatic scales. New instruments such as the saxophone were invented to smooth military band sounds. Violins of today are more powerful and have been modified from the original design of, say, Stradivari (even his instruments have been altered). Audiences are conditioned to both the new sounds and music from across the globe, so no matter how enthusiastic they are for early music, they can never hear it in the same way that it was heard and appreciated by the audiences for whom it was written. This is perhaps natural evolution of instruments and music, rather than technologically driven loss.

I believe that technology has actually been the motivating factor in the way music has expanded and developed. Technology has changed public taste and expectations of performance, which in turn has totally altered musical composition. This is quite contrary to the standard musicologists' views that art and literature were the driving factors. I am interested in how technology has shaped musical development and have discussed it in my book *Sounds of Music—The Impact of Technology on Musical Appreciation and Composition*.

10

Decay of Materials and Information Loss from Technology

Information and knowledge

In Chapter 9, I discussed how knowledge that is passed along orally is tenuous, as it needs a continuous chain of reliable intermediaries. It is equally susceptible to misinterpretation as languages evolve or are totally lost. Written records do not physically change with time, but problems of language evolution, nuances, and cultural significance of a text are just as serious. This is equally true of images, as we may have a 30,000-year-old cave painting, but we can only guess at the meaning. Nevertheless, written or electronically stored information is both valuable and interpretable for current information.

Therefore, in this chapter I will explore how different storage formats have survived. The historical changes and patterns they reveal are essential in giving confidence, or not, in our trend to relying solely on electronic systems. Modern electronic storage, whether on CDs, home computers, or distant central 'cloud' stores, now have immense capacity, and, since they are increasing at exponential rates, they could readily contain all previous records. Further, via Internet and other linkages we can, at least in principle, access them from any location. There is intense industrial marketing and media hype that this is the way we should proceed, and scrap older formats. Despite the advantages, I am also offering a forceful warning that a total move to electronic storage has a significant number of negative features. Certainly, in terms of home computer storage, many examples of data loss are already evident as progress and new operating systems and formats can all too often make older files inaccessible.

Similarly, there are dangers in relying on remote information storage, as we are dependent on electronic communications. Not only is there not universal electronic access, but in Chapter 1, I offered a plausible, indeed probable future scenario of major loss of satellite links and electrical power because of a solar mass ejection in a sunspot flare. That style of problem will be long-lasting and major. However, satellite failure or losses can occur for many other reasons, from age and decay of their components to more dramatic collisions between satellites. One such was in 2009 between the US Iridium 33 and the Russian Cosmos 2251 communication systems. Although the event only initially involved two satellites, this situation could easily worsen. An equally serious potential concern for both the USA and Russia is that if a military satellite were destroyed, it will be unclear if it were from a genuine accident or from destruction from another nation. The political consequences are potentially serious.

Because the orbits of most satellites are similar (the physics requires it), it is certain that the debris from each collision will destroy others at similar orbital distances from the earth. The key difficulty is predicting the life expectancy of the satellites.

This scenario could be used in disaster movies, but unfortunately, rather than just being fiction, it is based on the real possibility of collisions and destruction of satellites with fragments from earlier collisions. It is known as the Kessler syndrome. The problem is that satellites can fail, as did the Envisat, which is sitting as a large, lifeless, 8,000-kg (~8-ton) piece of expensive failed technology (it cost around $3 billion). It is travelling at up to 20,000 km per hour (~12,500 mph) in the same path zone where there are now some 20,000 fragments over 10 cm long travelling at similar speeds, plus maybe a 100 million fragments of less than 1 cm. Even tiny pieces at this speed have immense energy that can cause severe damage.

Appreciating the scale of a possible impact is difficult, but for those familiar with damage from high-speed rifle bullets, we can sense the scale of the collisions. In the low earth orbit region where most satellites operate, even tiny fragments can have 100 times the kinetic energy of impacting bullets. Larger items will scale up the collision energy by more than a thousand times. Not surprisingly, the data from 2014 indicate there is typically one satellite loss per year of the 2,000 or so satellites that are currently in similar geocentric orbits. Near misses between fragments and satellites, or other debris, are relatively frequent, and can be tracked by radar data. Actual impacts will just boost the number of fragments, and there is the ongoing concern that major losses of satellites could happen from debris cascades.

The tracking is essential, as the International Space Station now needs to be repositioned five to ten times a year to avoid predictable collisions with larger fragments. The only good news in this respect is that some satellites and fragments are moving in similar orbits and directions, and, at least in those examples, the relative fragment speeds are less.

Solar flare electrical power loss has happened several times, as have widespread power failures linked to overloads as climate has reduced the water in reservoirs behind hydroelectric systems. Water shortages have caused sustained electrical power loss. The electrical grid failure in three regions of India in July 2012 caused a prolonged blackout for some 620 million people (i.e. nearly 10 per cent of the world population).

Water shortages linked to power problems are equally likely in other countries. In the USA, difficulties have arisen from hydroelectric power from Lake Mead, plus closure of nuclear stations through lack of cooling water. We have a fragile and complex interlinked support system for our society, and loss of electronic communications may be one of the least critical items that could collapse. Nevertheless, it could remove vast quantities of stored data.

Central data storage can be compromised by terrorist and malware attacks. In 2016 there have been targeted sites which were overloaded by automated data requests that effectively closed the sites for several days on each occasion. More coordinated attacks could block general access to cloud stores over extended periods. So overall, it is important for us to explore and consider how permanent are our records, and also how easily we can refer to them. Quite independently, it is essential to ask if we can trust the information we find, and can we not only read it but also understand it. In all cases, from our own records to ancient historical items, some losses are inevitable, so in order to better preserve information, we must learn very rapidly why this happens. If we understand the causes of information loss, maybe (but only maybe) future generations will remember us and our thoughts. Finally, as I mentioned earlier, central storage may not continue to be at a low cost, nor free of government or criminal interventions, nor indeed is there any guarantee that it will it be updated in terms of electronic formats or maintained without ongoing funding.

Input technology and data loss

Our loss of written records is the result of many major factors, ranging from natural evolution of language, to politics and warfare, whereas other losses are simply that the materials used for records of writing and images have decayed. Slightly less obvious is that in each case a driving force for the losses is linked to improving technologies and globalization of the nations, particularly in warfare, and later in broadcasting, TV, and films. Despite the complexity and range of factors, I am going to show that there is a definite pattern for information decay and loss over the entire range of storage media, from skilfully and slowly carving a few words or pictures on stone, to typing on a computer (which is easy, fast, and immediate). The trade-off between these extremes is that the stone carvings can survive an incredibly long time (certainly many thousands of years), whereas my computer words are on a machine, in a software package and storage format that will undoubtedly be out of date in a few years. Unless I am very determined, my data will be unreadable in 20 years, except to computer-skilled archivists. The stone to computer pattern can be traced through all the other writing materials that have been used, and later I will offer a new 'law', or at least a suggestion that writing speed and storage survival time are not random factors, but appear to be linked in a semi-scientific pattern.

One factor that we rarely mention in discussions contrasting old and new writing technologies is the fundamental difference in the way most people approach their writing. For a stone carver, or a nineteenth-century scribe or bookkeeper, it was absolutely essential to stop and think very carefully before starting to carve or write, because once it was done it was unchangeable. For typing at a modern computer, I am probably typical in that I set out with a broad view of what I want to say, but am totally unconcerned with spelling errors (the software or I can fix them), and if I do not like my sequence of ideas, or have afterthoughts (very frequently both), then I cut and paste, as it is electronically trivial. I therefore wonder if the need to plan more carefully before starting, or the ability to make dramatic and frequent updates and change, has altered the way we compose and think. There is virtue in better mental planning, but I like the options to be able to correct and to modify.

Information loss from technology of materials

From politicians and scientists to the media, we are misled that technology is improving the world and everything is getting better. In the case of the ways we have written information (everything from shopping lists to love letters and science), this is definitely not correct. Indeed, the opposite is closer to the truth. The following examples will give a short catalogue of some of the writing media used over the last few thousand years and hint how long they might survive, and some of the reasons for their loss and decay. In every case, there is some trade-off between the ease and speed of writing and the length of time it could survive in that format.

Stone carving needs skill, strength, and patience, especially if it is intended for long-term survival, as then the choice of rocks will favour hard-wearing ones that do not readily weather and crumble. That means hard to carve. So carvings in granite have survived quite well, and there are many inscriptions on old buildings that are easily legible. The text is often on a large scale and limited content, but in favourable climates a carving on a building saying, 'This is the palace of the great King Og, ruler of the universe' can look impressive and last far beyond any memory of Mr Og. There are stone carvings from the earliest literate civilizations, so survival times of, say, 10,000 years are feasible. Aggressive climates and the use of softer stones to allow faster, cheaper carving clearly do not match this survival. A trip around any UK church graveyard will show some headstones are still legible after 100 or 200 years, but many others of the same age have been etched away in the weather, and eroded into a blur.

The obvious problem with stone carvings is that they are expensive to make, heavy, and not very portable, so are best suited for use at a fixed location—not items to be sent by post. Storage in kinder climates increases their chances of survival, as does being buried in a protective layer of sand. The case of the Rosetta Stone fragment indicates that it is possible to inscribe a lot of text on a material such as basalt and it is clearly legible a few thousand years later.

Considering stone carving is ideal, as it links writing speed and quantity of information, and shows that some examples survive, whereas a high proportion of stone tablets are lost, broken, or eroded. Overall, this means there is a typical 'half-life' where half the items existing at any one moment are likely to survive for this period of time. The concept of a half-life is familiar in science as the nuclei of radioactive elements (such as uranium or potassium) have a mixture of weights because they contain slightly different numbers of protons and neutrons (although they are chemically uranium or potassium). Some proton and neutron mixtures are unstable, and statistically half of them fall apart at a very precise rate. The time for the loss of half of the unstable ones is called the half-life. But the pattern of a half-life for survival is the same elsewhere, even if we are discussing writing on materials that gradually decay.

Stone is inanimate, so this view of life expectancy of materials is not the same as the way we think about human life expectancy. For humans reaching adulthood in the UK, we can normally estimate the percentage that may survive until 70 (the classic three score and ten years). Somewhat surprisingly, by the time we reach 90, the inanimate half-life pattern sets in. Survival statistics beyond 90 suggest for a group of 90-year-olds approximately half are likely to reach 91, and of those survivors 50 per cent reach 92, etc. To any 90-year-olds reading this, I am sorry to say this means only one in a thousand are likely to reach a century.

At this end phase, the model of a 'half-life' is close to that of the loss of information with an inanimate 'half-life', or of the decay of radioactive elements. Of the many materials used for writing, stone is clearly the best choice for long-term survival, as for hard stones a half-life of carved information may be a few thousand years. Faster writing became possible by incising lines into clay tablets and then letting them dry or be baked. This was the classic start of cuneiform writing. Survival was good, but the clay tablets were more easily broken or lost as they tended to be small portable items. Nevertheless, the concept of a half-life for clay tablet survival may still suggest a number of, say, a thousand years. Many more stone tablets were written than big stone carvings, so overall more may have survived (even with a shorter half-life).

By around 2000 BC, much faster writing techniques started to become possible by writing with inks on papyrus and paper. Bonuses included greater writing speed, light weight, ease of storage and transport, and for immediate usage, the materials would probably outlast a typical human lifetime of the period. For documents stored in dry and arid tombs in Egypt (inside protective stone boxes), many will have decayed, but overall a few have survived to the present day. The problems of failure are decay of the papyrus, attack by insects, fading of the inks, or destruction by dampness. Overall, this means carefully stored items may have a half-life of 1,000 or 2,000 years, but the general documents would vanish within 100 or so years.

The white-hot technology of writing materials moved to using animal skins, which have, confusingly and variously, been called vellum and parchment. As before, they were excellent for faster writing with inks and survival for transport, and, in good containers, a few have survived up to the present day. Fading of inks and destruction by bacteria and bugs probably sets a half-life at a few hundred years, but again, so many were written that a few have survived several half-lives and survived 1,000 years. Modern technology can slightly increase the legibility of the scripts that have faded, as often viewing them in infrared light offers a better contrast than for the original viewing of the ink in visible daylight. Vellum has not gone out of fashion, as the British Acts of Parliament are written on vellum precisely to offer long-term stability and legibility.

In 2016, there were proposals to move from vellum to an entirely electronic storage system. This change might have offered some initial speed advantage for the storage copy, but fortunately wisdom prevailed with the realization that electronic formats and software are highly transient. So access to an electronic primary store might only be measured in decades, not centuries. Electronic transmission and distribution of the laws is of course routine. (Later in the chapter, I will return to this topic with more comments on vellum and its importance for the Domesday Book.)

Examples of carefully preserved writing include political documents, such as the record of property owned by people in England after the Norman invasion. This 1086 book, called the Domesday Book, has been cherished, so it gives a distorted and extended view of the normal survival half-life of such writings. Other high-profile documents such as the Dead Sea Scrolls have been found (in very poor condition) from the first century BC. These are at the extreme end of many half-lives for the materials used.

The Dead Sea Scrolls are an interesting example of a technology with a short survival time, as the scrolls were in metal containers described as being made of bronze. Bronze (for most of us) is a hard alloy of copper and tin, which was widely used, corrodes to form a surface protective layer, and was valuable in sword making and armour. Bronze items have survived because of their protective coating. Bronze Age metallurgy was in fashion for more than 1,000 years. By contrast, the containers for the Dead Sea Scrolls were of arsenic bronze (i.e. copper and arsenic). It had some good properties and was probably no more difficult to make, but during the production the melt liberated large quantities of toxic arsenic vapour. The life expectancy of working in an arsenic bronze smithy was remarkably short, and it was highly unlikely to be a skill maintained in a family business. Arsenic bronze went out of fashion.

Modern style paper manufacture started in China around the first century BC, reached the Islamic world by the eighth century, and continued westward. The move from writing to various forms of printing on paper speeded up duplication of information transfer. Paper offers a good writing speed and a material that was more readily produced than parchment and vellum, but with a compromise of a shorter survival time. As with early items, the inks were also liable to fade or attack the paper. For example, many medieval black inks were made from acorn galls, which have rather high iron content. The iron gives a good black contrast, but chemically it attacks the paper. So the centres of loops (as in an o or p) could drop out. Many types of paper fade in sunlight or crumble. Because of the large quantity of printing (e.g. mass production of books), examples have survived up to the present. However, for one-off specific and normal documents, the average life expectancy had dropped to, say, a century or so. In the last century, typing paper damage could occur from the impact of the keys. Fading of the inks is still a problem, and much more recently many current warranties and guarantees printed on till receipts are valueless, as they will have faded long before the guarantee expires.

Writing information to computers

Early attempts at writing to store or send information, whether via cuneiform, pictograms, or later scripts, were all simplistic. The difficult part—interpretation and information processing—was made by a magnificently powerful and complex item, namely the brain. With numbers, various systems were used. For example, we could count up to ten using our fingers. Larger numbers were merely how many blocks of ten. For simplicity, adding a few key symbols reduced the number of marks that were required. So Roman numerals used symbols such as I, V, X, L, C, D, and M to define a unit set of 1, 5, 10, 50, 100, 500, and 1,000. The system is quite precise, but initially it was lacking a zero, and certainly it is not ideal for multiplication and division. Just try multiplying, say, 497 by 319 using only Roman numerals. Division is worse!

Many other possibilities exist, so instead of using blocks of 10 (from fingers) we could use just two (e.g. the two hands). This is called binary counting, so 0, 1, 2, 3, etc. become 0, 01, 10, 11, etc. Although this is not as easy to read for humans, it is actually far more practical for electronic computers, as the simpler electronic systems effectively have only two conditions: off or on (i.e. the 0 and 1 states).

Since computers lack intrinsic intelligence, we have to give them sets of instructions on how to make different operations, whether word processing or computation. This is highly tedious but, once instructed, they can perform calculations far faster than we can, so the time spent in the software writing is worthwhile. Instructions in terms of computer input therefore have to transform our letters and numbers into a code that is a mixture of on and off conditions. This is simple, but actually it applies to many other situations.

For example, in mechanical weaving looms, the machine has to be told to make, or not make, mechanical movements so as to create a pattern, or to change thread colour. These coded instructions were first made in the eighteenth century by Jacquard of Lyon in France. This technological advance speeded up weaving but, as in such advances, it changed the employment and habits of earlier generations, and the manual skills were lost at the expense of mass production.

The idea was improved by Herman Hollerith with a patent in 1889 for punched cards. Moving forward a century, it was seen to be compatible with computer technologies. So by the 1950s, computer inputs were encoded on punched cards in the style of the Hollerith cards. For the Hollerith cards, one wrote all the information that was to be processed as a pattern of holes on a stack of cards, and the software instructions were on another pack. It was slow, but so were the computers, which occupied vast areas with air conditioning to cool the vacuum valves (also called vacuum tubes). Rather more rapid access appeared with the advent of punched paper tape readers. Reading speeds improved, but corrections to typing errors on the tape were tedious. These writing methods were state-of-the-art but doomed to extinction; they each became outdated within ~20 years.

Replacements came in the form of magnetic tapes. Advantages included much higher feed speeds into the computer and easy corrections of errors; with care, an individual tape might last ten years. This is probably a generous estimate, as the tapes stretch and become corrupted if used frequently, so key data would be transferred onto new copies (copying errors exist but hopefully they are minor). Tape technology has advanced in materials, capacity, speed, and processing machines, and it is still used throughout music recording to make the master copies. It is also the norm for data storage in particle physics. This seems likely to continue, but the big caveat is that old tapes may no longer be readable, as the tape writing formats and sizes, as well as the tape readers, have evolved and changed. This is no different from saying cars have been around for 100 years, but the similarities between those in the early twentieth and twenty-first centuries are limited.

By 1984 (the year, not the George Orwell book), CD writing and data storage emerged, particularly designed for music to replace vinyl discs. To explain the reasons for this change, it is useful to look at the pattern of music recording. Most people will assume that music was not relevant for advancement of modern technology, but this is not true. It was actually the desire to record music that was a prime factor in the development of basic electronics. From this objective came microphones, amplifiers, and speakers, which enabled radio broadcasting, and all the subsequent expansions of all other types of electronics.

Successes and replacements of media for musical recordings

Advances in twentieth-century electronics revolutionized the world. Back in 1904, Ambrose Fleming (in the UK) invented a vacuum valve. The valve contained a hot cathode that emitted negatively charged electrons, which were attracted to a positively charged collector (the anode). This extremely simple concept meant that it was possible to have a device that allowed electricity to flow in only one direction. By 1907, Lee de Forest (in the USA) added a wire grid close to the cathode, so that small changes in the grid voltage caused large changes in the current transmitted across the valve. This was the basis of the first electronic amplifier.

The relevance for music was that in the nineteenth century, the attempts at recording had explored making impressions, first on wax drums, and then wax discs, by using sound vibrations to move a needle that scratched a pattern in the wax surface. However, this needed powerful sounds, and the playback was badly distorted.

Amplifiers of electrical signals offered a new dimension, as it became feasible to build sensitive microphones, and amplify their weak electrical signals for technologies ranging from loudspeakers to recording and storage on other materials. This meant one could make a master copy of the sound and duplicate it, rather than individually recording each copy. The disc only had a few minutes of recording time, but this was enough for dance music or songs. Gramophone music was therefore culturally very popular, even for non-musicians. It was an economic mass market. This drove electronic progress and eventually, by 1920, this had advanced to invent, and market, radio transmitters and receivers. Electronics had arrived.

The first electronic devices transformed the recording industry and brought music to a wide audience via shellac discs, but the playtime was brief, so there was the typical pattern of new variants driving the obsolescence and collapse of old-style systems and equipment. This was what electronics had done to the wax disc recordings in the previous generation. The record material moved from shellac to vinyl, and turntable speeds shifted from 78 to 45 or 33⅓ rpm. Hence playtimes steadily increased from 3 minutes per side to nearly 30 minutes. Each format dominated for, say, 20 years. Vinyl records were in competition with, or displaced by, magnetic tape, and tape became the preferred system by the early 1980s. Nevertheless, for classical symphonies and operas, etc. even longer playtimes were needed. This challenge was met by the introduction of the CD (compact disc). Reportedly it was made long enough to accommodate Beethoven's *Ninth Symphony* (about 75 minutes). The overall result was a collapse of the vinyl sales, which dropped to around 2 per cent of classical music sales by 2000. CD recording, giving 80 minutes of good sound quality, which had only appeared in the late 1980s, rapidly gained 80 per cent of the market by the mid-1990s. So within one decade, the music recording medium had changed.

CD storage

The large capacity of the CD for data storage moved CDs into the computer storage market. We should remember that in precisely the same period, there had been immense strides in computer power and availability. Therefore we need to ask both How long can information stored on a CD survive? and Will it become obsolete because of storage format changes, or from material failure? Therefore there may be different answers for rarely accessed CD stores, and those that are accessed repeatedly (e.g. as in music). The numbers differ, and this is important for survival of our precious music or computer-generated data. CDs are relatively rugged under normal handling, but they can become scratched, and the coatings are attacked by bacteria (especially in tropical climates). Decay of the polycarbonate coatings happens; the metalized surfaces can oxidize, and changes occur that makes them degrade with prolonged exposure to light or chemical vapours from the foam or plastics that were initially used in the packaging.

So, whilst carefully stored and unused discs might potentially last 50 or 100 years, those we are using on a regular basis have a far shorter lifetime set by climatic, chemical, mechanical, or bacteriological attack. A very obvious example is for music CDs in public music libraries, as these are invariably corrupted well within a decade.

In computer data storage terms, the extrapolation is that this may be equally relevant to the current CD, so data will be lost within a generation. For a system created in 1984, they have survived well. Music may again be the driving factor, or the villain, in ousting them, as there is the potential for improved CDs with, for example, interactive facilities to control tone balance or select different instrumental channels, etc. The concept is feasible, so it is highly likely to emerge—not least as they would generate a very large profitable market for the manufacture of the electronics needed to play them. Such a concept is commercially interesting, as it would revive a focus on CD-style formats that are now in competition with music downloaded in the form of MP3-type files. The MP3 versions are very compact and suitable for use with headphones, but for true music aficionados, the interactive CD, where one can modify balance of parts and adjust to local room acoustics, would be a revolutionary step forward.

Domesday Book—parchment success and electronic failure

The examples of the music industry—of changing formats for recordings—has emphasized that many successful new information storage techniques have had a limited survival time in prime position. It is equally instructive to recognize that the introduction of techniques which, at the time looked like a major leap forward, could collapse and vanish in a few years if they failed to anticipate the direction of technological development. A classic example for this evaporation of a new technology exists for the BBC's attempts to make a modern equivalent of the Domesday Book. This 1086 tome records details of people and property throughout the area conquered by the Normans. The aim was of course to raise the maximum tax money and legitimize their redistribution of wealth from the Saxons to the Normans. (They won the invasion battles, so this cannot be called theft.) The detailed lists of property and social arrangements also offer an incredible insight into life at that time. The maps and records were of considerable accuracy. It was a phenomenal and impressive document written in Latin covering 13,000 sites in Britain. Overall, it ran to more than 900 pages and 2 million words in two books. They have been carefully preserved to the present day. Interestingly, the final document appears to be in the handwriting of a single scribe.

To celebrate this event 900 years later, the BBC commissioned the production of some interactive videos with images and data collected throughout the UK. The project was conceived to offer a snapshot of modern Britain and was intended to be a valuable source of information for many years to come. In terms of the number of images and video clips, the challenge in 1986 was to find electronic writing systems, software, and hardware to record and access such a wealth of data. No standard media existed at that stage that could handle it, but the aim was to make it usable with BBC microcomputers, which had been introduced into schools by a government grant. Note that these state-of-the-1980s-art machines had no built-in storage disc, and the memory was typically 256 K!

Therefore an entirely new computer system was built for this purpose and the video material was stored on some extremely large discs. Unfortunately, the system was very expensive and government funding for such equipment to schools had ceased by the time the project was completed. The copies did not sell well, as the price in 1986 was roughly equivalent to that of a small car. Further, this entire process was clearly at the very forefront of 1986 technology. Unfortunately, it was not heading in the direction taken by mainstream developments, and by the mid-1990s there were thought to be no systems available to read it. Worse was that the computer formats were not compatible with 1990s technology, and the master magnetic storage tapes and discs had seriously decayed. Effectively the entire Domesday project had been lost in less than a decade.

In 2002, the matter was raised in the UK Parliament as an example of the dangers of keeping only electronic records, and an attempt was then made to retrieve the project. A working copy and a machine were found and the software deduced by a team who had worked in the area in the 1980s. The system was then transcribed to more current technology storage and display systems. The clear message from the example is that in a field of rapidly changing technology, even major projects can vanish within a decade. In this example, it was fortuitous that a recovery was possible as some of the people with the relevant technical skills were still alive.

Overall, whilst we generate data at ever-increasing rates, the survival time of the information has generally plummeted from thousands of years for clay tablets to much less than a decade for much of our computer storage.

Text, graphics, and photographic storage

If I return to information stored as written text, but now typed directly (or copied) onto a computer, then the pattern has moved completely away from using paper as the storage medium. Instead, it is sitting on a memory device, either in the computer or on some version of an external drive.

Unfortunately, we are now faced with two rapidly changing types of problem based on the technological advances. The first is that the software formats are developing and evolving so that our current word processing package has a lifetime of maybe five years before it has been upgraded or made obsolete. The second problem is that the format of the storage medium may equally become outdated. (I will discuss details a little later.) The net effect is that unless material is copied and transformed into new formats, it will become as unreadable as the Dead Sea Scrolls, but do so within a decade. This is a classic example of improved technology (computer power, operating software, upgrades and obsolescence of software packages, and storage formats) all contributing to produce an ever-shorter half-life for our documents.

A 'law' of the speed of written information loss

As a scientist, I gain confidence in an idea if I can see a definite pattern in the data, and even more so if there is a law that quantifiably describes what is happening, both now and in the future. There are also familiar examples of so-called laws that are actually just reporting trends. In electronics and computing, there is the famous 'Moore's law', which initially noted that the number of transistors on a computer chip was doubling every 24 months (with improvements, this speeded up to every 18 months).

Very similar patterns are cited for the improvements in other related technologies. Examples include the number of pixels per dollar on a CCD camera chip, or the amount of computer memory. All are increasing exponentially with time. For the CCD chip, the improvement rate has been around ten times per year, and for computer memory the value is nearer 100 times per year. In the technology of optical fibre communication, the number of signal channels, data rates, and distance are all important and interlinked. For fibre optics, the product (signal capacity × distance) has expanded by around ten times per four-year period. In this case the progress is not quite a smooth and steady advance, as often totally new technologies had to be introduced to maintain the expansion in signal capacity. It is interesting to see that the graphical plots of transmitted data rates also historically flow smoothly back to the nineteenth century, where Morse code telegraph or heliograph light pulses were used. The only (!) difference is that the transmission capacity has improved by a million million times.

When we consider the case of the survival of information written into different media, there are certainly trends running all the way from carving stone to writing on a computer. Finding an accurate predictive law is highly unlikely, as there are too many differences between the various materials, but a pattern does definitely exist. In my model, the trend I see is the following: The rate at which information is written and stored times the survival half-life is roughly constant (S). Imagine S as a measure of information survival.

Putting in numbers to test this model is certainly going to be contentious and a very personal matter. For instance, I can suggest that with a twentieth-century electric typewriter, a really good typist might have managed 80 pages per day for 250 days per year (i.e. 20,000 pages per year), and this text survived, say, 40 years before the paper and print faded away or were lost in filing systems and office renovations. Ignoring the funny units of 'pages' that I am using, this gives an S-value of 800,000. A comparison can be made with an excellent and prolific scribe who in 1086 produced the final version of the 900-page Domesday Book in about a year. The book has been preserved for some 900 years, which gives him an S-value of 810,000. The match may be deliberately biased, but broadly there is agreement in the effective range of S-values that emerge in such estimates.

One can take many other examples, such as the Rosetta Stone. There we can guess at the time taken for the carving, recognize it is far from complete as sections are missing, but the fragment is still in good condition. The arithmetic offers a comparable value of S. Similar numbers emerge for cuneiform written clay tablets.

For amusement, I tried using this S-value to predict the survival time of a typical doctoral thesis for a physicist, who has done computer calculations, and included computer graphics (i.e. where the computer power is equivalent to many years of manual calculation and plotting). The answer was a pitiful few months. However this is actually sensible! After submitting the thesis, there is a face-to-face examination (the *viva voce*) within a month or two. The best bits of the thesis are published, but the bound thesis volume then sits in a library, and may never be read again (or perhaps may be eaten by mice).

The modern phase in this trend is where powerful computers and Internet communication are transmitting immense quantities of information on short timescales. The implication is a very short survival lifetime for most of the transmissions. In fact, this is certainly true as most emails will only be read once. Hundreds are scams and junk mails that are immediately binned without being read, plus there are millions of blogs and tweets which litter the Internet, but which will rapidly fade into obscurity. In the latter case, Internet information loss is therefore not a totally negative feature. The real downside of the huge traffic flow is that we are simultaneously burying and losing data that is of value, and doing so ever more rapidly than in the past, when we were using other communication methods.

The very clear and crucial observation is that there is a trend that unequivocally demonstrates that moving to higher-speed writing and calculation techniques can generate information at a faster rate, but the material will survive for progressively shorter times before being lost or superseded. Further, the same pattern exists across a wide spectrum, from writing, computations, and mechanical equipment to, as I will now show, other types of information.

Pictorial information loss

I have already mentioned information encoded in the symbolism of art and the fact that, even for excellent-quality items, we may not recognize the symbolism implied or know that portraits are of particular significance. Art appreciation via classes, TV programmes, and books are now highly popular, but even for the modern experts there is a huge element of guesswork. When paintings are made, no one troubles to write down the currently understood implications, or to identify the mythology, politics, and personalities implied in the work. Art restoration has also confused many pictures, particularly if major repairs were needed. Further, since art by big-name artists sell for ridiculously high sums, there are an equally large number of copies and forgeries, which may date back to the time of the original work.

The modern equivalent range of difficulties has moved from paint or oil on canvas and paper to photography and computer-generated images. In the future, we will similarly lack the cultural or symbolic meaning of an image in the newer photographic or electronic formats. They may become electronically recorded items in obsolete software. The originals may no longer be accessible, or if the images have been transformed by updating into new formats, it will be impossible to know if they have been edited. There are many examples of political photographs where group photos have been edited to remove dissidents, or add in people who wished to be seen as supporting a regime. Family groups have similarly been emended for a variety of reasons, and from digital versions the original images are lost. Photographic electronic editing to improve, or change, the appearance of people is absolutely routine in many contexts.

Images, photography, and electronics

Art has served many purposes, from religious to political and social commentary, and with time, better materials, paints, or understanding of perspective, have all resulted in steady improvements (or at least from our present thinking). Nevertheless, technology caused a major disruption in this artistic development with the introduction of photography. It may variously be considered complementary, more accurate, or a totally new art form. Taking photographs became far faster than painting, but it also fits my pattern that the technological advance often produced images that have a far shorter life expectancy. In this case, oil paint portraits are likely to outlast photographic images (especially if these are in colour). Photography equally had a profound effect on painting styles: whereas early painting may have aimed for realism and accurate likenesses in portraits, photographs filled this niche so precisely that many artists moved into new directions that could not be compared with photographs.

Photographs offered rapid acquisition times, but the chemistry of the materials has never been as stable as for oil paints. As always, there have been trade-offs between speed of image collection, quality of image, and survival. The very early photographic attempts from the 1820s needed bright sunlight and a steady, non-moving subject. The process used glass-plate backing for the chemicals, but examples have survived nearly 200 years to the present day. Improved chemistry and processing offered finer detail and tone, but their long-term stability was often compromised, and their survival time suffered. Survival was also reduced through degradation of the papers on which they were printed.

By around 1900, progression to moving images was exciting, but individually each frame of such pictures is poor compared with the studio-made still images. Rather more troublesome was that for the moving pictures, the original cellulose backing was chemically and mechanically unstable. Not only could the film crumble, but sometimes it was a major fire hazard as it could spontaneously ignite. Effectively, such moving picture filmstrips had defined a chemically limited half-life for the images.

A later reason for the destruction of the film records arose with the new technology of having a soundtrack alongside the edge of the film. Film with sound instantly replaced the silent movies (and many of the stars who had looks but no screen speaking ability). The old stocks of silent film were destroyed, not just to minimize the fire hazard, but also because there is a valuable high silver content in the silver halide grains used in photography. The destruction was not limited to less popular works; even for fashionable silent films, with stars such as Charlie Chaplin, only a small number were retained.

By the 1940s, it was possible to make colour movies, and these ended the mainstream production of black-and-white films, both in movies and in home photographs. Most people see and appreciate colour, so this was a real advance, and the technology was desirable. Viewing in black and white is informative, but it has many subtle differences compared with colour. The apparent progress to colour was not without complications. The weaknesses of colour film soon emerged, as the chemistry is considerably more complex. Additionally, it has a lower sensitivity and it lacks long-term stability. For movies, prints, and transparencies, the colours fade and change with time. There is a very significant corruption of the original images, as all of us with old colour prints and slides will know only too well. Because the colours are far from stable, and they change and fade, they are actually worse than the more stable black-and-white pictures. Perhaps we are also more tolerant of fading in black-and-white or sepia images. The colour-changing situation is equally relevant for film-based home movies.

The next step in film technology was with instant printing cameras, initially for black-and-white pictures, but later versions had colour. These variously appeared towards the end of the twentieth century. They were briefly fashionable for, maybe, 10 or 15 years, but they still suffered from stability of the inbuilt chemistry. Therefore they were displaced by more reliable and cheaper electronic cameras. More recent toys include devices to offer small prints from the cameras.

The broad pattern in photography is as before: with introduction of better technologies, each new version survived for progressively shorter periods of time before being replaced by a new technology. The early pictures, obtained slowly and with limited information, were cherished, and there are examples surviving for almost 200 years; the half-life of silent moving pictures from the early 1900s was maybe 30 years (i.e. very few were retained because of fire or silver extraction); colour films and movies from the mid-twentieth century have distorted in terms of colour integrity, so that even if the films exist, much of the information was lost within, say, 10 to 20 years. This is even true for the film studio master copies, where they took care with the storage.

Home movies and videos went through parallel developments with cameras, processing, and projectors all improving. For new technologies, there are no initial standards, frame rates, or format, so videos were made in at least two major formats that were in competition. Each had some advantages, but neither enjoyed popularity for more than about 20 years before being displaced by electronic systems. Most family records on such home movies are stored in an attic and not viewed, as the playback equipment is no longer functional. Then they are disposed of as part of house clearance when the older generations die. So, all the jolly pictures of children, weddings, and parties are lost permanently within a generation.

Overall, in silver halide–based photography, the pattern is that information detail, colour, and survival are competitive factors, and the higher-grade colour images generated by later technology degrade more obviously and faster than the simple ancient black-and-white or sepia pictures on glass plates. In each version, usage rarely lasted more than about one generation, of, say, 25 years.

This is an absolutely ideal demonstration of better technology forcing shorter survival times for greater quantities of information, with quantitative data spread over two centuries.

A less obvious example of information loss comes from redesigns of computer processing. Just in my own experience, I have encountered many applications, such as the programmes for operation of dedicated research equipment, home packages for publishing, picture editing, or writing music. The processing was created as expensive software and written for computers that were state-of-the-art. In each example, the software was incompatible with the later, improved generations of computer processor. Rewriting was either not feasible or too expensive. I personally addressed the difficulty in my laboratory by keeping a variety of old computers, but this was a short-term solution with numerous associated difficulties.

Survival of electronic image storage

The demise of classical photography by the general public is mostly unrelated to the weaknesses or chemical instability of the photographic process, but instead to the ability to have electronically recorded storage systems with CCD cameras. Less obvious is that film photography required many skills in actually taking pictures with the correct exposure and focus, plus a major skill in the film processing, since any error at that stage was irreversible. Similarly, editing and printing of the film images needed highly professional methods, which meant a poor photograph remained a poor photograph for ever.

This contrasts with images taken and stored electronically. Because of the potentially immense and lucrative market, the number of CCD pixels that define an image has increased remarkably fast and, even more surprisingly, the price per camera has fallen steeply. This has been matched by great improvements in the quality of the lenses, even on mobile phones. The advantages of a modern CCD camera are very obvious: they include compactness, good colour fidelity, high resolution, high speed, time lapse, an enormous number of pictures stored in the inbuilt memory, plus the ability to instantly review the attempt and, if necessary, take another picture. Video and high-speed action shots, short bursts of images, plus electronic auto-focus and anti-shake features all add up to a camera package that, for the general user, is far more convenient and superior to silver halide photography.

There are further advantages that with subsequent processing of the digital image, it is easier to change colour balance, or edit pictures to modify them. Critically, the changes can be made on electronic copies, so the original is still available. Professional photographers will still find some preferable features of the photographic process, but even they will routinely use electronic CCD cameras, not least as modern detectors can have more than 30 megapixels. This gives a resolution formerly obtainable only on high-grade film.

The one-generation pattern is again evident, as within less than, say, 10 years of high-definition CCD cameras appearing, their sales are falling, as most snapshot photos are now taken on mobile phones. They are instant and the pictures can be immediately sent to friends and, as predicted by my survival 'law', viewed once and never seen again, or erased within hours.

In my own photographic and CCD records, I am a typical camera user: I progressed from film through several generations of CCD camera. The earlier CCD quality now looks so poor that I have deleted many of the images. Equally, of the later ones that I have kept, there is an overload of pictures on my storage system, so rather than having a few special photos of key events in my life that I might revisit, I have hundreds of images of minimal long-term interest, so I look at all of them far less. In terms of information survival, the average picture lifetime and viewing has rapidly fallen. This is precisely the message of this chapter. More means less.

The mobile phone cameras extend the pattern. Users of mobile phone cameras are not extreme if they take 30 pictures per day when with friends, which equates to around 10,000 pictures per year. If they receive an equivalent number from their friends, and spend as little as 2 minutes or so per picture in terms of taking, sending, receiving, and viewing, then this is consuming an hour per day of their waking life. Phrased differently, it suggests that many CCD pictures have a viewing lifetime of minutes, at most. The associated negative factor is that they are clogging the Internet and mobile phone signal capacity.

Computer power and information loss

The exponential growth in computing power, information storage, and electronic communication via phone and Internet has created a transformation over the last 25 years that was previously unimaginable. We are so enthralled by the new toy that it has dominated our lives and we have put total faith in it. Our enthusiasm is frequently well founded, as it has opened rapid communication across the world, faster calculations, and better weather prediction.

Despite the good features, negative aspects also soon became apparent. For example, it has revolutionized the way in which we shop, but in so doing it has reduced the support for local community shops and local employment. More hidden from us when we make Internet purchases is whether the taxation on the profits from our purchases is still arriving to help our own government or indeed any government (or is the seller based in an offshore tax system?). Isolation of sellers and companies from the community isolates them both physically and mentally, and they then lack the associated interest and support in the nation to which they are selling. They are definitely not encouraging local industries. Instead, it is driving globalization of products with consequent rises in bulk transport, even of goods as simple as garden produce when it is locally out of season. These factors all have negative features, since they cost fuel and power and in turn are depleting natural resources and contributing to global warming and shortages of water in many countries. Not least is that the pattern increases unemployment in many types of job in the more expensive countries, such as the UK or the USA.

Less apparent at this stage is whether, with greater computing power, it will become feasible to have far better robotics in manufacture. Robotic technology will raise profits for the manufacturers, but match this with ever more human unemployment. The argument that there will be more leisure time is false; in reality it often means unemployment, and it completely ignores the satisfaction of personally working and achieving something. For myself, I see reduced prices that raise unemployment as an unacceptable trade.

We have been totally overcome and obsessed with the advantages of high-speed electronics and communication, but, assuming we hope to survive the next 25 years, then we need to rapidly understand the side effects and long-term consequences.

Patterns in data storage

My focus in this chapter is on how technological advances are causing ever more dramatic problems in the survival of information and data storage. Therefore it is interesting to contrast how information was stored in the pre-computer era and at present. Analyses of data storage in 1986 show it was totally different in character from the present day.

First, virtually the only method was with analogue systems. Of the total data stored each year, it was estimated that it was made up of roughly 25 per cent for music on vinyl discs and cassettes; ~13 per cent for photographs on print or negatives; and ~60 per cent on videocassette format. Since then, the quantity stored has exponentially grown by at least a factor of ten every decade (i.e. by the present day, that is a thousand-fold increase—and still rising). Most obvious is that the pattern has dramatically shifted so that there is probably no more than, say, 5 per cent still in the analogue format: digital tapes come in at ~10 per cent; DVD, etc. at ~20 per cent; and hard disk stores ~40 per cent. In other words, we are totally relying on the computer-accessible data stores.

Over this period, the computer storage formats went through a variety of disc and tape formats with ever-increasing size from, say, the 250 kilobyte floppy discs up to the current terabyte storage devices. For those who are embarrassed to ask, the prefixes kilo-, mega-, and tera- just mean changes from a thousand, to a million, to a million million. Very roughly, the range of disc storage from the 1980s to the present is equivalent to a scaling from an area the size of the top of a golf tee to an area the size of five football pitches. (I find this easier to imagine than just the numbers.) The progress is impressive, but the only easy way forward has been to make all the former methods obsolete. Typically, each style of storage medium has vanished within a decade, as the size and power increased.

The overall effect is that few of us have equipment that can read any of the information that was stored on the early systems. This is not a passing phase, but a problem that will be ongoing as it is the only sustainable way to continue the expansion of data storage.

Many companies advertise that because of this difficulty, the solution is to keep all our data on files stored in some vast remote site, euphemistically called 'the cloud'. The benefits are clear that *if* the cloud management could continuously update the formats of our stored data, then it might avoid obsolescent equipment and inaccessibility. However, this would imply the cloud company had access to our files, with several clear disadvantages: (a) the files may be confidential; (b) the files may be encrypted; (c) the files may be corrupted in the process; (d) experience in running the cloud stores is limited, as the concept has not been in existence for long. Therefore, we have no way of guessing which cloud companies will survive, or, if they are taken over, whether the existing arrangements and contracts will continue. Another potential difficulty is that if we are paying for the storage (and eventually this will be the case, even if the present loss leader is to make it free), then when we die, or our company goes bankrupt, our payments will stop and the entire store may be deleted, or at least be inaccessible without payment.

There have already been court cases where people wished to share data in their cloud (e.g. in one case, a vast music collection), but the company not only refused this but implied it would not be available after the death of the owner. So it is not family silver being passed from one generation to next, nor the bundle of love letters tied in a ribbon that can move to the next generation (and historians). Instead, death may result in the loss of all items in the cloud. Thus the cloud is an excellent step in terms of storage space, but a disaster in every other respect. If your will and bonds are stored in the same way, then they could equally vanish into the clouds.

The other very significant fact is that for cloud storage, we must be able to access it, so in times of power loss, or Internet traffic jams, this will not be feasible. The final problem is that, as will be discussed in the chapter on computer crime and cyberwar, a determined government, terrorist, or misanthrope could destroy a great deal of data with each attack. Judging from all other trends in this area, then the destructive power of such attacks is also increasing exponentially.

The same level of caution is needed with respect to other suggestions for a paperless society; for example, the suggestions that UK medical records could be totally in this format.

I am perhaps being overcritical of cloud storage, as for many companies it has the benefit that it can be accessed by many employees at different locations. Nevertheless, I personally would ensure that somewhere I had a total back-up copy, which is secure and accessible only to a few, and isolated from email and Internet access. For everyone else, the moral seems to be that one should only use it if you could survive the loss of any information stored there.

A broader canvas of the half-life concept

Perhaps less obvious is the concept that secrecy can also have a limited life. For non-written items known only to a single person, then the 'half-life' is set by the death of the person with the information. When secret information or documents are spread across several people, then the possibility of disclosure increases. We can make attempts to keep information secret, or restrict access, and need to do so for a diversity of topics. The reasons are various and span political, commercial, and criminal activities (plus combinations of them). The most highly classified top secrets may be so restricted that even the people with access may not fully understand the implications (e.g. politicians may not understand the real dangers from new weaponry, or the military may not have considered agricultural consequences of a new biological weapon, etc.).

Many industrial companies also need to maintain secrecy of their production methods and ingredients. So secrecy is necessary, but the word will mean different things to different people. However, the speed with which information can now be accessed and disseminated via electronic communication means that any security leak (of whatever type, and for whatever motivation) means data that were once firmly locked away for decades can now suddenly be public knowledge. This is not a passing phase, as even 'top-secret' information means more than one person will have access to it. Therefore a single whistle blower, a person of conscience, or someone keen to disrupt an organization can distribute the data or ideas. In terms of half-life of secure information, technology has potentially moved us from, say, the 50-year rule of non-disclosure to instant worldwide distribution.

From many aspects, this may be desirable, as widespread dissemination can reveal criminal activities of organizations who are claiming to be acting in the national good, or charities (or their recipients) that misspend their funds. It also has revealed criminal activities that had previously been covered up by religious or other large public and private organizations. Therefore the perpetrators can no longer guarantee they will have immunity throughout their lives. The converse is that totally open access to all types of knowledge and information would undermine commercial activities and security (both local and national). I doubt that we will ever have the ideal balance, but undoubtedly technology has opened the gates for those who wish to make public disclosures. Whenever the storage computing systems are linked to other equipment and networks, then the potential for hacking to read or alter the information is incredibly high.

What are the possible solutions for data retention?

I suspect the difficulties of retaining and accessing information need different approaches depending on the type of material that we want to keep. For personal files, documents, financial records, family photos, etc. we will certainly need access during our own lifetime and some material we may wish to keep for future generations. Here we hit the standard dilemma of how to deal with dynamically changing software and hardware. Invariably the new systems are incompatible with the old, so, unless we act whilst we can, the records will be lost permanently. Therefore we need back-up copies that are not accessible to malware and hackers, and ensure it is updated into the latest versions of software and storage.

For really critical items, keep a hard printed copy. This is not an exclusive approach, and if we are confident with cloud storage systems (or their next generation), then they may be a secondary line of defence. Their weakness is that changing formats may make our precious cloud documents and photos obsolete unless we are accessing them to make the updates. Since this implies a very large market opportunity, it seems inevitable that companies will emerge that will address programme and format obsolescence in order to offer an update service, but unless they start soon, our older material may already be dead.

We should not personally feel inadequate with this problem, as it is equally valid for major institutions, companies, and research centres. An often-cited example is that NASA has archival retrieval problems of data gathered by space probes. Certainly it is not information that can be regenerated. In their case, the situation is made worse by the fact that space missions take a very long time to plan and prepare, as well as running for many years after the launch. Wisely, they opt for reliable and well-tested equipment and software. Unfortunately, this means that in computing terms, it is out of date by the launch.

I have mentioned the evolution of music and video storage systems, and these do not just apply to home items and CD collections, but also to the radio and TV broadcasts and films. This again is not a problem for individuals, but for major institutions and national archivists. The problem is certainly understood. In the UK, major archives of recordings are held by the British Library. Hidden in their repositories are some 2 million recordings ranging over all the sound formats from wax cylinders through all the later variants of magnetic tapes, records, and CDs. Many are at serious risk just from mechanical decay, whilst others need equipment to play them, and most would benefit from transfer to a truly stable electronic platform. This is a pressing challenge, not least as the quantity of new material is rising rapidly.

The next category is one that epitomizes my rule that the faster information is produced, the shorter is the survival time. Here I am referring to all the electronic signals generated by Twitter, Facebook, Instagram, blogs, and Web pages, which are sometimes merely personal, but in other cases offer useful information that is worth retaining. Such material may never have been in print, so we are faced with a storage that is in a purely electronic format. Again, the reality is challenging; the current estimate of this exponentially growing database is that by 2014, more information was produced by such routes than had ever been written in all the recorded history of the world. Storage capacity is increasing, but routes to cataloguing and accessing it are extremely hard to perfect, so many of the desirable items may just become hidden and lost, even if they are actually stored somewhere.

To offer some perspective for the scale of the electronic storage problem, we should recognize that camera technology, both as separate items and those built into phones, allow such rapid electronic image capture that most of the images are of very temporary importance. Fortunately, this means we discard them and have no real need to keep them for the long term. Many images we send from our phones are meant to be just a 30-second comment or entertainment. This fits neatly into the scheme that the faster data are generated, then the shorter is their survival time. If we took a different view on life and wanted to keep all such images, then the storage problem would be phenomenal. Present estimates suggest that across the world we take around one trillion (a million million) pictures annually, and the rate is increasing.

The other purely electronic information that is generated by governmental and national bodies now includes vast quantities of data, reports, legal documents, announcements, and social comment that have never existed in a paper format. The loss that happens when their websites, etc. are closed or blocked means we are losing cultural, educational, and legal documents and records at an ever-increasing rate. Because we are currently losing this storage battle, the British Library has used the phrase of a 'digital black hole'.

If we are making data stores via electronics, then, even if the storage medium does not need continuous power, we will still need electrical power in order to access the data. This is a weakness in the storage chain, not just because formats alter with time, but because we may need our stored information at times of a major crisis. My Chapter 1 scenario of a minor natural disaster that even temporarily removes electrical power says that at just these critical moments the data are inaccessible. Prolonged power loss is worse because the life expectancy of all electronic data storage, and the items stored there, items could suddenly be reduced to zero. An equally possible vision of 'data doom' would be terrorist or governmental interference (local or foreign) that wipes out national archival storage. In my view, judging from the examples of malicious or politically motivated corruption of electronic data and computer systems, then this is actually far more probable that the rarer natural catastrophes.

The final question—will we be remembered?

Storage systems and software are inevitably going to be dynamic systems that evolve and advance, not necessarily with total compatibility with earlier versions. Therefore for any records, photographs, and images, or other items that we want to preserve for our families or posterity, we need to ensure that whichever storage medium we are using, we keep updating the information. The reality is that in terms of a desire to be remembered, we need to admit that we will rapidly fade into obscurity. Many people construct family trees running back several hundred years and these may have names and dates, but in fact they say virtually nothing about the people. A few may have been famous (or notorious), but the bulk of the billions who preceded us have totally vanished from our records. Therefore, my conclusion is to forget about vanity; instead, enjoy any present fame and family that we have, and admit that within a generation or so we will be lucky to be remembered by anything more than a name on an ancestry tree or on a gravestone.

At the more pragmatic level of how storage systems have evolved and information has decayed, I will offer an analogy: we walked before we had bicycles, trains, cars, aeroplanes, and public transport. The newer technologies may be faster, but feet are well tested and reliable. So records on paper should not be despised.

Unequivocally, the message is that high-speed production and delivery of stored information is matched by a shorter time that it will be accessible. This is not some new idea, but it applies to all the information systems we can consider. Any teacher will confirm that a rapid delivery of even a simple topic will not be understood if it is presented too fast, but a slow and relaxed delivery will be understood and remembered. This is equally true for romance and sex.

11

Technology, the New Frontier for Crime and Terror

How to succeed in crime with minimal risk

The downsides of being a professional criminal is that the profits may not be large, evasion of the law is variable (unless you are very successful), and there may be dangers both when committing offences and also conflicts with other local criminals. An intelligent potential criminal will therefore diligently study science, technology, and computer software. This will offer a safe and profitable career path. Technology offers a dream opportunity as, first, the crime can be committed on people or businesses in a foreign country, and second, with a wise choice of country, there will be minimal chance of being extradited. This is matched by virtually no personal danger from the victims, and via the Internet the entire world is the target. Depending on the level of computer skill and intelligence, a whole range of options suddenly become available. Not surprisingly, the actual result has been a phenomenally rapid growth rate of cyber and associated electronic crime.

Most of us, the trusting general public, will be aware of computer scams that may start with a phone call or email claiming to be the bank, tax office, or a software company. Whichever route is used, one aim is to gain access to our computer; once there, it is possible to read the contents. The available data provide passwords to bank accounts and credit cards. With less technical skill, the cybercriminal can attack with a wide range of scams where goods are on offer, which appear to be from reputable companies, but instead our payment details result in a bank transfer to a foreign country, and our money disappears.

My examples are just the obvious methods, but in fact these scams are only the tip of the iceberg. In terms of overall hard cash, cybercrime is extremely significant. For example, in the USA, the total money lost by such routes was estimated to be around $1 billion in 2013. This estimate is increasing; it had doubled by 2015. Of more concern is that the scale and sophistication of the methods for such crimes continue to increase. So caution and better security are essential. Once money is lost, the chances of any recovery are small.

At a personal level, we are likely to be aware of our own bank transactions. But for small sums, which, for example, purport to be a standing order for some charity or insurance premium, we may believe we have just forgotten that we set it up. For the criminal, such small sums are not very tempting, so larger numbers per theft are preferred, or a multiplicity of smaller scams. However, most of us are likely to rapidly note the loss of a few thousand pounds. This is a slight dilemma for the criminals, as it may mean individuals are less likely to be worthwhile targets for major criminal organizations—not least as we will complain, and this rapidly triggers criminal investigations.

The intelligent criminal therefore wants both large sums and a system that is lax in spotting errors. Hence the ideal target may well involve complex multinational organizations or even banks where there are many foreign transfers taking place. A classic example emerged in early 2015 when the security company Kaspersky investigated losses from cash machines. The initial problem appeared to be that some ATM sites had been remotely directed to release money at a selected time, without the need for any banking card. It then rapidly emerged that this was just part of a much wider and more extensive cybercrime network, which had been running for several years at perhaps $1 billion per year.

Rather than merely hacking into the system, one method was to use video images with cameras within banks. These revealed details of bank processing methods and procedures. Duplication of the activities then moved money between banks; for example, by the SWIFT transfer route. In some cases, the source account total had been accessed and briefly altered to read, say, 20,000 instead of 2,000. A transfer of 18,000 meant the owner of the account was unaware of any change in the balance, which had only appeared to increase during a temporary hiccup of a few milliseconds. Equally, the SWIFT mechanism was fooled, as the source appeared to have had sufficient funds for the transfer. Once the foreign 'account' had accumulated sufficient funds, it was closed—after the money had been removed.

The psychological skill was never to target the same bank for too large a sum, so as to avoid detection and criminal investigation. Total losses to a single bank were kept typically below £10 million over one or two years. Ten million is a very large number for most of us, but it is tiny in terms of an annual bank turnover; indeed, the loss is probably no more than the annual salary of the directors.

There are many banks, so there are many potential targets. Kaspersky's first estimates were that prior to 2015, several hundred banks from across at least 25 countries had been robbed (or are being robbed). The total sum so far is measured in terms of billions of dollars. In 2016 a variant of the method removed around $81 million from the Bangladesh central bank, in part by blocking the printing details of the transfer. A parallel effort to remove $951 million from the Bangladesh account in the Federal reserve bank of New York only failed because the transfer details had a spelling error.

The other critical feature for such international criminals is that, over the past few years, the number of the new multi-millionaire criminal fraternity who have been successfully prosecuted and funds regained is around zero.

These commercial losses are not to be confused with stock market and bank losses caused by automated trading, which estimates trends and futures. These are equally the result of technological and software 'progress' where the computer predictive models are flawed, and are designed to be independent of human vigilance and control. To investors, the losses can be substantial and, except in legal terms, may also seem to be totally criminal, lax, and incompetent on the part of the fund managers.

Should banks consider a return to electric typewriters and handwritten ledgers?

Smaller-scale computer crimes

Crime generates an equally profitable business for honest cyber experts and consultants, as many websites offer and sell advice on protective measures such as anti-virus software, and also detail the various weak points in our defences where attacks are likely to originate. Social networks are particularly dangerous for the unwary, as they involve people we are communicating with, who initially may seem to be a new 'friend'. Initially we trust them. The 'phishing' scams and information gained about details of our lives, or even incriminating photographs, may soon follow. The latter are used for blackmail, and this is a growing and very profitable crime. Blackmail scams are mostly hidden, but there has been an increase in the number of suicides they cause. Overall, the

UK 2013 estimates ranked the relative danger for access to these crimes by social networks in Google, LinkedIn, Myspace, Twitter, and Facebook from 5–39 per cent (in the order listed).

People effectively are trusting, gullible, and put themselves at risk—the expert advice from Government Communications Headquarters (GCHQ) (the UK department concerned with security and electronic espionage, etc.) suggests that 80 per cent of cybercrime could be blocked or avoided by better security and risk management, and a less trusting attitude to strangers whom we do not really know and may never meet (not forgetting that the pictures we see of them may be of totally different people).

The UK scale of cybercrime

Cybercrime is not limited to activities of unknown people gaining access to details of a bank account, etc. Indeed, many examples will involve people we know and work colleagues. Like charity, cybercrime can begin at home. Many industries, town councils, or charities will have large amounts of money flowing through their system, and this money is handled by a wide range of employees. Access to such financial dealings is rarely tightly limited, especially in large organizations, so 'standard' cybercrimes include invention of fictitious employees who are paid salaries; double salary payments to employees using more than one address and bank account; or purchase of non-existent goods and services. Many of the same criminal problems already existed in paper-based accounting, but the electronic versions are far simpler to generate and often less traceable to the perpetrator.

Major companies will have many hundreds of employees who will know how to make relevant cash transfer details and passwords. Some will be careless, and others may accidentally contribute to the crime. Similarly, in competitive industries, the theft of knowledge and production information or marketing strategies, etc. will all provide profit to those who can extract the data from their competitors. These thefts of information (not money) are crimes that are hard to pinpoint, and unlikely to be prosecuted. The scale is worrying as it is growing.

A 2013 government report estimated that the UK economy was annually losing (at the very least!) some £27 billion to such crimes. Their breakdown spread the pain as follows: £21 billion to business, £2.2 billion to government, and £3.1 billion to private individuals. If GCHQ is correct, then there is a potential for an enormous improvement. The numbers clearly show why it is attracting not just the standard criminal, but a totally new class of thief who feels secure and untouchable in such actions. In some ways, the crime is committed because the thief may never see the victim and therefore has no remorse.

These sums are in fact paltry compared with the money hidden in tax havens. Estimates vary on how much is not being used to pay taxes to home countries, but 2013 numbers suggest that at least £100 billion was lost in tax revenue, and globally towards £5 trillion was stored in tax havens in foreign countries. The sum may exceed the total national debts of nations across the world. These sums are so immense, and doing nothing to reduce poverty or improve the life of most of the world's population, that in my view the very existence of such tax havens is unquestionably criminal. I am sure I am not alone in also saying that the various governments around the world who lack the integrity, courage, or ability to tackle this issue are equally guilty. In my context of technologically or electronically aided crime, such tax evasion is definitely a leading candidate. They are in fact very obvious, as many of the 'banks' for the deposits may not even exist in the form of a building, but are purely electronic sites.

One often-cited example is the Cayman Islands, where there are nearly 400 registered banks, but only around two dozen carry on business with residents and non-residents of the islands. For a population of less than 60,000 in this 100-square-mile island group, the density of banks is excessive.

I am surprised that there are no effective 'Robin Hood'–type hackers targeting this immense wealth and redistributing it to poorer nations or worthwhile causes.

Hacking and security

Software and security are written by individuals working in small teams. For most products, they have target dates by which to deliver and, in a competitive industrial market, the aim is to finish rapidly. If their efforts appear to work, then inevitably they move on to the next project. Rarely will everything they planned be 100 per cent successful, either because they made errors, or the customers want items in the package that they had not considered or included. This is obvious, as with most software packages there are regular updates and replacement versions. The mentality of such software programmers is that they are trying to create, not destroy, therefore by instinct they are not the best placed to consider all possible loopholes and weaknesses in security. This produces a running battle: when security is breached, a solution has to be found. Sometimes systems can run for 20 years before a weakness is noticed and exploited, and this is particularly problematic as it is unlikely that the original programmers will still be functioning in that role, or that they would remember the details of their approach. It may even be that advances in computational power have allowed a weakness to be exploited that was not feasible in the original plan. A perfect security defence is therefore a nice ideal, but totally unrealistic.

Looking at major hacking events reported to the public merely shows that defences of major companies and government agencies are equally viable targets for those who are skilled and determined. In terms of espionage and political disruption, the major nations are all actively operating at the criminal end of hackers (perhaps they believe they are justified in terms of national security). Their efforts may not merely be to gain knowledge, but to sow items of disinformation, change details in documents of other nations or companies, or cause damage and destruction. For example, a software bug introduced into the operation of a Middle Eastern centrifuge used to separate uranium isotopes destroyed the units and set back the programme of developing nuclear power (and possibly weapons). Whilst no nation claimed responsibility, unofficial reports suggest that the attack was organized by maybe half a dozen programmers working for a few months (i.e. a very minor-scale activity).

A 2014 example of access to a Sony Pictures studio main computer indicated the scale of the problems that can occur. The studio was due to release a controversial film comedy about North Korea, and this triggered a politically motivated attack at a more sophisticated level than might have been expected by the industry. After copying the data, the malware destroyed the servers and computers, and wiped them clean, apparently with relative ease. For many companies this could mean total destruction, but it indicates just how vulnerable major industries are, and how little defence exists for individual users. It is probably a reasonable assumption that government computer hackers were involved, and equally that many governments have gained access to computer networks of other nations. The only difference is that in most cases the intruders are likely to be sitting as passive listeners, rather that exposing their electronic moles. The listener approach means they have access to planning and, in the event of a conflict, can then destroy enemy communications from within. The spying methods are more subtle than the Big Brother scenario of *1984* but more intrusive than could have been imagined at the time of writing it, since the events were set 30 years in the future, before the potential of computers and electronic communications had been appreciated.

An example of a very simple politically driven intrusion into our electronic communications was made at the end of 2015, when a concerted attack was made with a modest number of automatically dialling smartphones. The activity spike this generated was some ten times greater than all the other Internet traffic. Overloading specific targets has also been noted on numerous occasions, both before and since that example. The real concern here is that such actions may only be practice exercises prior to a wider-scale attempt to cause the Internet to crash from a general overload.

I have also read that modern missile guidance systems are now so complex that the entire control chip is often fabricated in one country, which need not be the one that plans to include it in their missiles. The opportunity to have software sleeping in the chip that could redirect the missile if it were used against a target friendly to the chip maker is all too obvious to me. The situation seems remarkable, if true. However, the precedent of a NATO country selling missiles to South America, without changing the technology that informs radar signals it is a NATO missile, caused considerable damage to a NATO power when the missiles were used against them (i.e. they assumed it was not targeted at them). So the later missile coding story is also feasible.

Understanding the technology one is using is absolutely essential. A related missile example occurred during the use of a guided missile attack. The problem appears to have been that the satnav coordinates were set for the enemy target, but the operator did not notice a low-battery warning on the device. The voltage fell lower and the technology of the missile then automatically reset the coordinates to the local site. Once the delivery system was activated, it then accurately exploded at the new location, and destroyed the command post.

Effective anti-hacking

The only secure defence against electronic, Internet-connected hacking is not to be connected. Therefore in an industrial company (or national facility), an isolated stand-alone computer system is more secure than one that communicates to other computers of the same company. At first sight, this seems a perfect solution, but for many years it has been possible to record keyboard typing via sensors unconnected to the computer. Such sensors sitting in range of the computer controls will therefore broadcast passwords and input to a more distant receiver. These are familiar espionage units that have been used to record bank activities from outside of a building. Unless the system is in an electronically screened room, and powered with a truly isolating power unit (i.e. not just plugged into the mains supply), then computer isolation is not 100 per cent guaranteed.

An interesting approach was taken in the Kremlin where it was reported that some 50,000 euros were spent on electronic typewriters for truly important secure work. I assume these are only used in bug-free, screened rooms. The Kremlin is skilled in such activities; there was a report that they once built a foreign embassy with local workers, and managed to install excellent surveillance of the major foreign power, who thought they had saved money by using local labour.

Equally surprising was the disclosure that the president of the USA does not carry a mobile (cell) phone. The reasons were not specified, but intelligent guesses may be that it would be feasible to track his exact location at all times, intercept messages, or even use the phone covertly to listen in to conversations.

Espionage and security

In terms of method, there is probably little difference between governmental espionage and criminal access to private or company data. Governments can of course operate on a grander scale, and some have implemented searches for keywords in every email, Facebook, or other document that is being transmitted via the Internet. The same mass probing is equally feasible for landlines and mobile phone communications. These are major and relentless intrusions into personal privacy. The difficulty from the security viewpoint is not detecting key emotive words (i.e. this is simple character recognition), but to recognize the context and prioritize documents or voice messages that reveal criminal and political activities. The total traffic volume is immense, so the separation of the threats from the rest of the signals is extremely difficult and must be largely automated.

Our images of technology in espionage, whether international, industrial, or personal, have undoubtedly been influenced by imaginative scriptwriters, as for the James Bond movies, and many other films and TV programmes. Surprisingly, many of the toys used in the spy films did not exist at the time of writing, but have since appeared in reality (e.g. a laser cutting through a steel plate, wristwatch communication, and head-up display spectacles). My instinct is that once some new gimmick has been conceived, then, unless it breaks the rules of physics, it can be built. If we invest enough time, effort, and money, we are likely to find a solution. As a physicist, I have set problems to university tutorial groups to guess how a particular system might work.

The psychology here is interesting. If I posed the question about a device that did not exist (and said so), then there would be very feeble attempts at suggestions. However, the very same group of students would be imaginative and enthusiastic in generating new ideas when I said the problem or equipment already existed. Knowing that a problem has been solved gives us confidence to produce our own solution. Novelty and the need for an original idea are often too challenging.

The true scale of scientific progress is difficult to appreciate and trust, except in an area where one is knowledgeable. This is because magazines, papers, and TV science programmes inevitably provide superficial comments for a mass audience. They try to attract and hold our attention, so may imply excessive claims for a new discovery or product. For example, they reiterate how much more powerful are computers or other electronics compared with those of, say, ten years ago. Of course, they are correct but, since we cannot see how many pixels are on the CCD camera, or physically relate to the speed of a computer chip, it is hard for us to fully appreciate the technological advances.

Offering numbers to say the improvement is by a thousand million times means it is 'a lot'. The magnitude is difficult or impossible, to grasp, as we have no familiarity with very large numbers. Effectively, we become unimpressed with the impact of the true scale of the improvements. Personally, I need some tangible physical scale of the changes that have taken place. Even as a scientist, I have trouble; something as basic as saying that in an atom the diameter of the central nucleus is 100,000 times smaller than the overall diameter gives me no real feel to the scale, even though I believe it to be true. However, if I see the same factor presented as being the width of a blade of grass compared with the size of a football pitch, then I have a tangible impression of their relative sizes. (I still need to visualize a sphere of that size!)

So to underline the changes in performance and size of electrical equipment within living memory, I will compare the radio transmission equipment used by British agents dropped into France at the start of the Second World War with more modern systems. State-of-the-art 'portable' radio transmitters in 1942 had a range of nearly 500 miles, and they neatly fitted into a suitcase, which weighed around 30 pounds (~15 kg). Of course, they also needed a 6-volt car battery to power them, and it was necessary to set up many metres of aerial, and then carefully tune the transmitter frequency. This offered Morse code transmission, not direct speech. In order to change frequency, so as to avoid detection during a long transmission, it was necessary to physically alter some of the components. Not too surprisingly, detection of an early secret agent, with a transmitter system, was not difficult, and the average life expectancy before detection (and execution) of the agent was a few weeks. This latter fact was not emphasized when recruiting volunteers.

A 'miniature' state-of-the-art version, at 20 pounds, was developed in 1943. Whilst such instruments steadily became better, even by the 1950s a clandestine radio transmitter was still a large briefcase size, and weighed about 20 lbs (~9 kg). It ran off the electrical mains and had a speech channel as well as a Morse code transmitter, but the range had increased to maybe 2,000 miles. Our daily use of smart mobile phones with instant picture transmission (and the option of encryption), clearly underlines the technological progress in electronics. Indeed, modern mobile phones not only have more computing power (and speed) than the highly cited Enigma decoding computer of the 1940s, but also are thousands of times more powerful than the NASA systems for Apollo launches.

Similar advances in cameras, remote sensors, and bugs for spying are now readily available. Even small drones can be used to fly and spy, or be used in agriculture to assess when plants are ready for harvesting. In the case of vineyards, the sensors can pinpoint precisely which vine should be harvested. (The onboard infrared detectors monitor the sugar content of the grapes.)

In terms of espionage, or intrusion into the life of others, I suspect that the drones that are so valuable in agriculture and are used for military surveillance will feature much more in future discussions. Their price is decreasing, whilst their performance is improving. This is classic technological advancement. Therefore, their range of applications will increase, and this is bound to include criminal activities as well as merely looking into the grounds of other people's properties.

We may assume that spying on us in our homes is difficult, but in addition to transmitters that can be built into devices left covertly in our rooms (e.g. which externally appear to be an adaptor plug or telephone socket), a good spy no longer needs to enter the room to leave the detectors. Some hackers have used malware to access our computer cameras and microphones, and do so without activating the normal display lights that indicate the camera or microphone is running. Equivalent bugs in mobile phones can allow transmission when we think the phone is off. Basically any spying scenario that we can envisage is likely to become feasible.

If you are leading a lifestyle involving sensitive conversations and contacts, and your mobile phone battery frequently needs recharging, then check for surveillance bugs.

Technological progress is fantastic, but so are the opportunities for misuse and abuse.

Mobile phones, cars, and homes

Mobile phones offer communication, photography, and Internet access that we all use. What is not apparent to the user is the additional information that allows the receiver and transmitter stations to pinpoint the location of the caller and the person receiving the message. We do not need to be using the phone to disclose our location. This is highly valuable in tracking a lost child or vulnerable adult, but it in turn means anyone able to access the records can plot our movements and details of whom we have talked to. This is an undesirable lack of privacy. Far worse is that we now have many applications that allow us to remotely access equipment at home. These include the ability to turn on the heating, draw the curtains, start the oven, and so on—all these functions are operated via the phones. Performing precisely the same tasks can be initiated by anyone else who has managed to log into our phone security. The gimmick of having electronically controlled remote door locks for our homes and garages now seems far less desirable, and presumably may (or should) violate insurance conditions.

In the case of cars with phone and security access, a number of other features have emerged; some were predictable, but others were unexpected. There has been a UK trend to dispense with keys in the more luxury range of cars, and instead of a key, use an electronic key fob to open and close the doors. A little foresight would have spotted that this causes problems if either the car battery, or the fob battery, drop in voltage.

To overcome the problems, and also to enable garage servicing, access to the onboard car computer has been offered via a simple security override system that is available to garages. Inevitably these same devices, or alternatives, have become used by criminals to reset the fob code, and so gain entry to the car in order to steal it, or items in the car. Some insurance companies have now limited coverage on such key fob luxury cars. A return to large crook locks on the steering wheel is now preferable, as most car thieves will just move on to the next, easier target.

The onboard car computer control systems record many details about engine performance and automatically make adjustments to optimize it whilst the car is running. In the case of the highest-level super cars, they appear to do everything else as well. This is fantastic progress. As drivers, we are unaware of this background electronic control, except that we are impressed that we no longer skid and drift when going too fast around a corner, and can accelerate and brake without skidding—the computer chip controls the power steering.

The computer is rather more skilled than this, as it may monitor or change tyre pressure, release the airbag, and transmit GPS location and speed. The sensor responses can be used to assess the skill and ability of the driver (this information is now used by some insurance companies, which offer lower premiums for good driving, especially for young drivers who otherwise would have very high premiums). Additionally, it has mobile phone communication connections and can transmit (and receive) any of the data to other sites. In the event of a crash, it can automatically call for help and give the location. Poor driving is recorded; the computer data have been used for convictions of speeding, etc. Distance sensors help in automatic parallel parking, and on variable cruise control they can speed up or slow down the car if the surrounding traffic changes speed. Overall, this means our skilled driving is not totally under our control. Instead, we have delegated some of it to the onboard computer, or, more worryingly, to anyone who can access the computer.

The design features were all made with the good of the driver in mind, but the downside is that precisely these features can be exploited by a potential criminal. In practice, it is feasible to hack into the system, or add a bug via the CD player, so track the movements of the target person in the car. It then only is a small step to override the driving instructions via the car computer, to fire the airbag, or cause the car to swerve, change tyre pressure, or accelerate the cruise control or steering to cause an accident. Perhaps in some future film plot the murder will have been committed by someone in a different country by remotely causing a fatal accident. Detection and proof will need an imaginative and computer literate detective.

Interestingly, I considered the possibility of two of the preceding examples (the SWIFT transfer and car control) and wondered if I should include them, as they might be used in criminal activity. I can now relax because both have been used, demonstrated, and have been reported. A demonstration of taking command of the car by controlling speed, door locks, radio, steering, etc. was made, and the consequent panic was a recall of 1.4 million Fiat Chrysler vehicles. Presumably this is not a unique example, but it emphasizes that those working closely on a project may not be able to take an outsider's view and sense such potential problems. The designers are focussed on their product performance and may not be able to mentally step back and consider how it could be misused. Nevertheless, the film plot option, or reality, of gaining insurance money or causing a car accident has not yet appeared.

In 2016 similar demonstrations were made on controlling aircraft and overriding the pilot controls. Enjoy your next flight.

I have recognized several other potential criminal opportunities, but will reserve most of my thoughts on them until they appear to have been used, and have been openly reported.

One potentially illegal feature I had not considered was to use the computer to modify the response of the car engine during emission tests. This has powerful marketing implications—a factor not lost on the manufacturers. In 2015 there was a disclosure that one major car maker had taken this route to falsify the performance, and hence emissions, during the official testing. It may not be surprising, but the initial disclosures admit this has been the case for more than 1 million vehicles, and later data then increased the number to 11 million. It now appears to have been more widely used by many major companies. In countries such as the UK, where the test results define both sales and the road tax, it means a distortion of sales and a major loss of income to the government. There also appear to be deliberate distortions of emission data with cars with two- or four-wheel drive. The UK test is made solely using the front wheels. Under these conditions, the computer controls can minimize emission, but once out on the road, and converted to real-world four-wheel drive, the computer software restrictions are lifted on some models. Performance changes, so sales increase, but so do the emissions, by an extremely large factor (more than ten times in one case).

One further oddity of ideas, similar to that of the motor industry, is the attempt to dispense not just with car keys, but also keys to control house entry systems and garage doors. The technological advance is a sophisticated voice recognition system. The fail-safe mode is not to open the door if the power supply fails (or rechargeable battery is flat). That means you are locked out of your home. Another unavoidable failure is that coming home with a bad cold, or drunk, may mean your voice is imprecise, and the door sensor will not recognize you. Equally, listening to TV impersonators, it is clear that their ability to mimic a person is so good that they will easily fool any system that is mass-produced at low cost.

Medical records

Producing and maintaining a database of medical records has very obvious benefits, not only for people who need medical treatment when away from home, but also because the system can offer a mine of information on regional practices, types of illness and disease, and the performance of the different practitioners and hospitals. All these are extremely valuable in planning and improving healthcare across the nation. Open access to such records is far less desirable if the data are used to obtain personal information about insurance coverage, job applications, or the lives of those listed on the databank. There is therefore a delicate balance between the potentially useful and negative access to the information. Purely in terms of healthcare, the medical histories are supposedly confidential in terms of actual name, so the health service assumes they have limited detailed access by some 10,000 personnel.

This implies an extremely low level of real security, as within any group as large as 10,000 people, there will be many who are careless, overly inquisitive, or willing to use their access for illegal purposes. The UK level of criminality may not be excessive, and perhaps health employees are more caring than most people. Nevertheless, even for extreme criminal acts, such as murder, from a group as large as 10,000 we might typically expect (and detect) at least one murder per year. So for less extreme crimes, there will inevitably be intentional misuse of the data. This is reality and therefore unavoidable. We cannot have both total security and a detailed database with widespread access.

A study in progress in 2016 in the USA has been looking at the more sophisticated ways data can be mined from records of doctors' prescriptions, pharmaceutical sales, and medical records. Currently these have been encoded to preserve patient anonymity, but they still inevitably include factors such as age and gender, and in the USA there is a fragment of the zip code of the patient, and of the medical practice, as well as links to previous medical history. These generalized data are very valuable for looking at patterns of drug usage and localized diseases and how they spread, and to determine if there are links to cultural, economic, or industrial activities. They also disclose if some medical practices are abnormal in their prescription patterns. Finally, the records offer valuable knowledge for both the pharmaceutical industry and government health agencies.

Nevertheless, each of us has a unique medical history; knowing, age, gender and general locality of patient and doctor means that it is now feasible to access much wider databases to identify precisely whose records are being presented. This is clearly simple in a small township where there may not be many people of a particular age and a particular medical history. In reality, it is becoming feasible even for people in larger cities. The downside of such disclosures may be undesirable for a wide range of reasons. However, the reality of confidential health-related information has ended. This is a factor that may open opportunities to increase insurance premiums, or even be used in blackmail.

Security is further weakened as data can be densely packed on a CD or pen drive (or be sitting in a computer), and there are many examples admitted where such items have been lost or stolen, together with vast amounts of confidential medical information. The information can then enter the criminal, or public, domain.

A second example of a double-edged sword is the acquisition and storage of DNA information. Again the data could readily be taken at birth and stored nationally. It would offer great insight into genetic diseases and changes in health problems, or local anomalies that may be caused by environmental problems. The genetic identification would also be of value for forensic applications. Nevertheless, access to the precise names and locations of the people identified by their DNA requires an extremely secure and well-controlled system, operated by a minimum number of people. Whilst regional data may be widely used to help understand the effects of the environment on disease, individual identification would need to be far more tightly controlled.

Even for forensic purposes, the first step of matching the DNA of a person who is not on a criminal register should be channelled through a secure and encrypted route (i.e. not open to any member of the law enforcement). Release of the name or other details would be at a much higher security level than is used for medical records. In many cases, there should be no need for specific identification, but abuse of the data would be attractive not only to insurance companies and employers, but also to criminals. The opportunities for misuse are clear.

One area where this is particularly obvious is that numerous genetic studies have confirmed that, whilst maternity is a fact, paternity may be a matter of opinion. For the UK, the current studies suggest as many as 2 to 4 per cent (i.e. up to 1 in 25) of fathers may not be the biological parent of all their children. Similar, or higher, rates are cited from many other countries. Consequently, automatic access to such data, rather than analyses made during a paternity dispute, could undermine many families, or be used as leverage in blackmail.

A final comment on DNA data is that the public, juries, and police usage has tended towards total acceptance of DNA evidence. Reality has now indicated that corruption of such evidence can occur, and many borderline cases have resulted in errors of prosecution. It is a salutary warning that total confidence in any new technologies is unwise.

Distortions of existing data

Our reliance on Internet access to information means that, once we have chosen our keywords, the search engines give a list of options, and we invariably head for those at the top of the list. They therefore gain a higher ranking and rise even further up the list. I doubt that many people will search beyond a few pages of references for general enquiries, so the later items gently vanish into obscurity. Again I suspect that as a scientist my approach may be a little different from general users of search engines. When I try to find an article related to an obscure science problem, I will often persevere, and search down maybe 20 or more pages of potential articles. This can be rewarding, but I would never do this for a general-interest problem.

Overall, it means that it is possible to influence which items are seen near the top of the list. It is then a short step to influence the list, and those items in conflict with the views of the manipulator become buried. In many ways, this is worse than the pre-Internet situation, as formerly a library search was not prejudiced or easily influenced. It just needed patience and dedication. Overall, I find that the more extensive Internet searches tend to reveal cross-referencing and perpetuation of misquotations and errors. The party game of Chinese whispers, where a message becomes modified as it is passed on along a line of people, is quite similar.

Direct and intentional copying without revealing the source is plagiarism, and it a familiar practice using Internet searches (especially effective if taken from items appearing on later pages of the search list). Student essays generated by Internet searches will inevitably contain some plagiarism, and are criticized unless skilfully rephrased. Academic supervisors also extensively plagiarize, but invariably use many sources, and this of course is called research. Feelings about being plagiarized are mixed. I once found a book with an entire chapter lifted exactly from an article I had written (but the source was not cited). To some extent, I felt pleased that the writer had been so discerning! On another occasion I met a colleague who said he very much liked one of my science books, and even had an electronic copy on his pen drive. I did not know such an electronic version existed, but apparently it had been copied by his government for open use in their national laboratories. He gave me a copy!

Internet data are often updated or they may be deleted. It is also often quite difficult to guess at the accuracy of a source, even if it appears to come from a respectable establishment, rather than from a personal opinion or blog. In this sense the data and 'facts' are very different from written records in archives and old newspapers. In the pre-electronic cases, we may have been able to make a better value judgement as to the accuracy and possible bias.

Recent disquieting reports include the removal of material from the Internet because it cast influential people in a bad light. There was no question that they had acted criminally or in unsocial ways. However, if they were now rich and influential, then they claimed that they had the 'right' for previous acts and statements to be forgotten. This is a totally unacceptable premise; deletion should only be permitted if the original data were false, not to improve political or criminal image. The fact that the examples of deleted information only seem to be related to prominent and rich individuals is significant.

On social networks (e.g. Facebook), there is much information that is personal, false, or malicious, and it seems highly reasonable that there should be redress and data removed. But for genuine news stories, any selective removal, to please the affluent or powerful, is precisely the same rewriting of history that is undertaken by the victors in a war, or a dictatorial regime.

A final word of caution about sourced material is that it may not be accurate—not because the original was in error, but because in some intermediate repetition it has been modified. This is true not just of text, where, for example, there may have been translations and copying, but also of photographs. Glossy magazines abound with examples where an original photograph has been 'touched up' or blemishes have been airbrushed out. Better technology has made this distortion of the truth far simpler. As many skilled photographers know, it is now possible to modify electronic images down to the pixel level. Generally the aims are to improve the aesthetic view of the image, to slim a stomach, add more hair to a bald head, or include people who could not attend a group photograph, etc. The adage that a photograph cannot tell a lie is now totally wrong, and one must be cautious in viewing modern electronic images. The same skills in modifying photographs and electronic documents are of course criminally and politically exploited.

Confidence in security systems is strongly dependent on who might benefit by accessing it. If there are strong political or criminal motives to access, destroy, or alter stored records, then security will always be compromised. Secure systems are devised by humans. So if one group can write the relevant software, then an equally skilled group can break into it. Such a concern of potentially weak security was revealed in a disclosure that police surveillance-camera data were being backed up in commercial 'cloud' storage facilities. Such data may be legitimately accessed by many police departments, but inevitably this means that they cannot be very securely encrypted. So although 'cloud' stores may be excellent for personal items, it is open to abuse and alteration. In major criminal investigations, the camera evidence justifies greater protection.

I have mentioned encryption as adding a barrier in the defences of stored information and communication. Currently this can be reasonably well achieved as the computer power needed to break the encryption is limited to major computing systems and expertise. So for personal security, most governments can break encrypted signals. Higher barriers and better security may exist for government transmissions. In a book focussing on advancing technology, I should mention that one computing technique that has been in development for many years is termed quantum computing. It is hard to achieve and only very simple demonstrations have so far been successful. However, there is a real possibility that at some future date it will become feasible. Once this happens, it will be possible to break any of the currently used encryption codes.

Technology and terrorism

Warfare, invasions, expanding territory and empires, forced religious conversions, and mass destruction of those who disagree with us are apparently part of human nature, as they have been documented throughout recorded history. We have obviously learnt very little over the last few thousand years, and although the quality of life has improved for many people, the deep-rooted problems continue. As part of our technological advances, there have been major efforts to make better weaponry and more effective ways to kill and control our fellow humans. I am clearly making some mistake in condemning such progress, as all governments fund it, and we never openly suggest the politicians and weapons manufacturers are interested in power and wealth rather than the good of the nation and humanity. The same weaponry and techniques developed for national defence and supremacy are of course readily accessible to all parts of society. Rather than being used for our benefit, they can rapidly become the tools of the trade for crime and terrorism. From the viewpoint of technological progress, the advances are significant, but they imply that the threats to the stability of society are advancing out of control. Terrorism is a very simple activity. It has no rules, no Geneva Convention guidelines; in the case of devout fundamentalists, if the terrorists happen to die in the process, it may be seen as a bonus and a direct route to heaven.

This was precisely the message that was used to activate the Christian Crusades against the Holy Land in the Middle Ages. The justification to the soldiers who went to fight and be killed was that it was a just and holy war, and death would be rewarded in the afterlife in heaven. The fact that the Crusades were politically motivated via a so-called religious justification is absolutely incredible, since it was from a religion that has an original concept of tolerance to other people. Both sides in the conflicts said much the same, so maybe this was just an extremely evil and perverse aspect of the Middle Ages.

Not all participants were so naive: although less widely reported, many of the leaders were making substantial fortunes from the pillage and the land they stole during the wars. Indeed, some crusade expeditions were more honest about what they were doing; for example, one fleet from Venice changed course to sack cities such as Constantinople.

The differences between modern terrorism and that of a few centuries ago is that rather than merely kill a few people in an explosion or an armed raid, with potential danger to the attackers, the weaponry and potential of modern bombs, etc. allows one person to kill thousands quite indiscriminately, and do so remotely. The cyber routes that allow a skilled solitary person such power is extremely worrying, as it may be used in a totally random event. Additionally, as seen by the New York Twin Towers' terrorism, the actual weapons need not be advanced military equipment, but in that case just commercial aeroplanes.

Such 'progress' is setting a trend that is difficult to combat. The far more significant concern is that the mentality of terrorism is still rooted in the basic weapons approach of the knife, gun, or bomb. It has rarely evolved into understanding and using the chemical or biological possibilities that could be far more destructive. The reasons may be that such weapons do not have the immediate publicity impact that terrorists generally demand in order to further their cause, or that they may realize that once activated, the progress of, say, a plague or other disease may be uncontrollable, and will equally attack the terrorists' community as much as the target population. These are considerations that may be ignored by extremists or psychopaths, so I suspect we have not yet seen the truly dark side of technology-driven terrorism.

The future

The few brief examples I offered here are disquieting, as major beneficiaries of advancing technology are the criminal fraternity and terrorists. Computer, Internet, and electronic crime is a boom industry that is growing exponentially. The profits are large, detection and prosecution are currently difficult, and such types of crime will continue to grow. All aspects of society, from individuals to government agencies, therefore need extreme caution. For me, a particularly perturbing feature is that, whilst reading and thinking about this chapter, I could envisage many more potential criminal uses and opportunities than have been openly cited or discussed. For obvious reasons, I have not presented them. Neither am I yet planning a career change, but I see why there is the temptation for cyber-related crime. It may be driven by greed, political, religious, or malicious behaviour, but it is undoubtedly an aspect of our lives that is here to stay.

12

Technology-Driven Social Isolation

Isolation driven by technology

Technological advances are inevitable, often desirable and beneficial, but inevitably acquired at a cost. There will always be many people who cannot follow, understand, or cope with the new version of the world, and for them technology may mean social isolation, inconvenience, and unemployment. In the following sections, I will give examples of how perfectly good technology impacts and isolates various groups of people. This is despite the fact that the changes are often promoted as being of benefit to the very same people who are struggling. Good intentions and reality do not always go together.

It is easiest to consider three different age groups, starting with those at school or teenagers; the next bracket includes those in mid-life who expect to be working; and the final group I will euphemistically call elderly. Actual age definitions depend too much on physical and mental ability to be more precise. There are also very large differences between the rich and poor, and the region or nationality where people live. In the UK, for example, there are strong regional differences for health, wealth, and life expectancy. These are only partially linked to technology.

For all of us, our first hurdle was to start with caring, intelligent, rich parents. If we were fortunate in that respect, then from on it was probably easy to move into to a well-paid career. These social advantages will not only offer a comfortable lifestyle but also increase our life expectancy. Indeed, there is a simple pattern that progressively higher average income, for similar jobs, results in progressively longer life expectancy and better health. Money is not the only factor, as those with independent control of their lives will survive noticeably longer than equally paid people who were totally subordinate in their work and life. The differences can be as much as five years. Diets, local attitudes, and genetics all play a role. Whilst it is possible to look at trends, there will be many people who do not fit into the broad patterns.

There are similar difficulties in estimating how much influence and change in society is occurring as a result of technological progress. It is possible to compare different groups of people, and monitor factors such as life expectancy, income, physical fitness, and health. These are often closely interlinked. They may be confused by changing lifestyles and the overall state of both the country and the world.

Therefore government or academic studies will frequently draw very different, or even totally opposed, conclusions. The benefits of technology will totally dominate the perspectives of some, whereas others will recognize that benefits for some are invariably bought at a cost to others. Here I am taking a narrow view of looking at difficulties that result in social isolation. In terms of numbers, perhaps only 10 or 20 per cent of the population are financially, physically, or socially disadvantaged, or actually suffering from the various new technologies. However, in the UK, this means real people on the scale of 5 to 10 million (i.e. a number equivalent to the entire population of Greater London). Therefore my concerns are real, and should not be ignored as being just a problem for a few per cent of the country. I am using the UK as a model, but clearly the same factors apply worldwide.

The three age groups I will discuss each contain roughly one third of the population. Young means up to the end of the teenage years, the middle group are those likely to be involved with work and employment (including employment as housewives or other home-based roles). 'Elderly' is a rather emotive term and a matter of attitude as well as age, but, at least for the UK, this probably mean those who are retired, so are older than their mid-60s. This is a positive view, as in some countries 'elderly' technically starts at 40 (or less). Even now, many people will remember when 40 was termed 'middle-aged'; conversely, based on their activities and mental attitudes, many in their 70s do not yet think of themselves as elderly. Concepts of life expectancy have also changed. For example, in Hong Kong, those born in 2016 are projected to reach 84, so many will reach the twenty-second century.

Isolating technologies and the young

For the young, the primary villain in my accounting scheme is our fixation on continuous communications with mobile phones and computers. This is not real human contact, as the electronics do not provide tones of voice that distinguish between threats, affection, irony, humour, or puns, any of which might have been implied with the same set of words. Therefore, misunderstanding can easily be triggered by prejudice, a misreading of the text, or reading into it what we want to hear.

As I have already mentioned, these are all features that are crucial in the design of robots, where we must recognize the problems in order to make them more marketable. However, we gloss over them when dealing with humans. For us, electronic contacts totally lack the crucial and highly valuable information and subtlety that comes from seeing facial and body language, or responding to pheromones, or all the other chemical signals that we unconsciously exude during a face-to-face meeting. Instead, the machines give us an anodyne and imperfect conversation via a device that has been consciously designed so that it does not even provide a true power and frequency response of the words and sounds. That is feasible, but apparently we find it too disconcerting. So, electronic conversations mean that, at best, we are having only a partial contact, even though we can do it from anywhere where we have phone coverage.

Despite this failing, the current generations have a pressing need to contact people who are not actually with them. This obsession frequently seems more important than talking to those around them. The pattern is immediately obvious in cafes, where one can see a whole table of 'friends' using their mobiles for talking, texting, or emailing other people.

Vocal, visual, and physical contacts are not just adult human needs, but absolutely essential between parents and babies to form family bonds and stimulate mental development (both for children and parents). Unfortunately, electronic communications can destroy this life-defining situation, in the very early moments when they should be central to family bonding. It may be unintentional, but we start to isolate babies and small children when the adults are having mobile phone contacts with friends. It is very common to see babies in prams and pushchairs, with parents who are wheeling them along, totally ignoring them, as they are chatting on the mobile.

The early months of childhood are absolutely critical: this is a time when conversation, eye contact, and touch between parents and child are life-forming. Without it, not only social but also mental skills can be permanently impaired. Once this opportunity is lost, it can never be re-established. The losers are therefore not just the children, but also the parents and indeed the entire society, as we will be raising children, and a new generation, who basically feel unwanted, unloved, and inadequate, and unable to properly communicate. Their future parenting skills will similarly be undermined.

I suspect that one of the reasons the parent-to-baby isolation has increased in the last century is that old-fashioned prams had the baby facing the pusher. In 'modern' designs of pushchair, however, it is fashionable to have the baby facing forward, away from the adult, with no visual contact between them. It seems there is the need to either return to the earlier pram design or add some 'new' technology in the form of a rear-view mirror onto modern buggies. Obviously this is too simplistic to be marketable, so instead let me suggest perhaps a pair of screens and cameras so baby and pusher can see one another!

It is very easy to underestimate the importance of the personal contacts compared with conversations transmitted solely by other means. There can easily be errors and ambiguity when reading the words that would not exist if we were listening to them with all the inflexions of speech.

Written ambiguity can of course be both useful and intentional. On one occasion, I had a reference for an applicant for a job with me that said, 'If you can get him to work for you, then you will be very lucky'. A telephone call clearly revealed what the referee was implying. I employed someone else.

This same loss of non-ambiguous range of meanings, when we have no access to the speech patterns, delivery, pauses, intensity, inflexions, tones, or speed of delivery, is a major but overlooked difficulty in our daily electronic correspondence via texts and emails. The written communications are also inferior to personal contacts, even if the text has been carefully considered. We prefer speech. In truly high-technology developments of future personal robots, there is now a detailed study and emphasis on trying to make robots that respond to our moods, not just a set of commands. This means they will recognize changes in the pitch of our voice, delivery speed and volume, and the way we pick our words. For example, when addressing a small child, a friend, someone we dislike or are jealous of, or an adult in authority, we consciously vary all these factors. Once this robot speech recognition becomes routinely possible, then the irony is that we may then prefer to have a conversation with a sympathetic robot (or electronic 'friend') rather than an electronic conversation with a real person. Indeed, since the robot will inevitably be only programmed to be caring or sycophantic, we will become ever more isolated from one another.

There are very clear reasons for this, as we can be sure that the electronic friend or robot is never critical, offensive, or lying to us. Therefore our sense of personal worth will be inflated, and we will always be happy with them. It is a very false sense of reality, but for many people it will be effectively a drug-forming habit, as it will continuously boost our self-image and release serotonin and other happy responses in our brains. Commercial success in robot manufacture will follow, but the value for individuals is less obvious.

As children grow older and go to school, the situation may not improve. Large numbers of children are now given mobile phones at an early age. Undoubtedly, phones are useful, but dependence on them is setting the pattern of electronic addiction for communication and further undermining their ability to make person-to-person contacts. Financially, there will be peer pressure to have the newest version of each phone (i.e. a major annual expense), with all the currently fashionable apps and games, plus home computers, and yet more electronic games. Those who do not have the latest gadgets will be treated as outcasts. The world of children at school is a harsh one.

Young people are far less secure and worldly-wise than adults, so are unprepared for the pitfalls and misleading effects of the electronic age of communication. These come in various guises. The first is that if they are not in frequent contact with many other children, then they feel they are isolated and rejected. The number is more important than the quality of the contacts, and in many cases the electronic contacts would never be matched by genuine friendships if they spent time together. There are further dark sides to these networks, because children (and adults) naively put random thoughts on their Facebook pages, or post anonymous comments that may appear quite differently from the way they intended them, or be spontaneous, and something they would not have said if thinking carefully. These widely spread comments may be false, or out of context, but once sent, there is no way anyone can alter them, as the first impression has the impact, not the retraction or correction.

The same applies to cyberbullying and rumour-mongering. These are genuinely major and serious problems, in part because they can be anonymous, and in part because the impact of the comments may not be understood by those making them. Even more importantly, they are serious because they are often comments that the blogger would not have had the courage, or stupidity, to say in a face-to-face meeting. Children are even more sensitive than adults, so these are deeply emotive and damaging aspects of electronic communications. Cyberbullying is widespread, permanently damaging, and difficult to block.

Teenage children seem particularly vulnerable to friendships developed electronically, without realizing that the new 'friend' may not be anything like the image that is sent electronically. Photos and other details can be totally fictitious. Further, there have been numerous examples where young people have sent indiscreet or pornographic images of themselves to their electronic friends. There are criminals encouraging such pictures, as they are then used as blackmail to avoid disclosure. The mental pressure on the children has resulted in a number of suicides. The numbers and total scale of such exploitation and criminal activity is difficult to estimate, but appears to be increasing.

In total, the time spent on the mobile or at a computer screen (or watching TV) is soaring, with a reported doubling within the last decade. The many detailed surveys from both the UK and the USA are consistently confirming that the number of hours is increasing steadily, especially among children and young adults. This isolation from real person-to-person contacts undermines the longer-term social abilities of everyone involved, but is an increasingly divisive trap for those who are insecure, vulnerable, or from families that offer no alternative sense of direction. In many cases, the total time spent on phones, online, or watching TV, is quoted at incredible numbers of six to ten hours per day! Many children say they go to school to sleep, as during the night they are constantly checking their networking pages to see if they have been mentioned.

Unfortunately, this is not the end of the list of bad side effects. Defining life in terms of a computer screen or mobile phone means we invariably adopt a bad posture (most of us do), which undermines health. Eye strain, lack of exercise, and remaining indoors are therefore follow-on problems, as well as poor eating habits. This has caused a significant rise in myopia (short sightedness) within the twenty-first century. Numbers cited are increases of around 75 per cent in both Western Europe and North America. It is not surprising that within the last decade, the physical ability of the young, compared with their parents at the same age, has shown the first-ever reversal that the younger generation are not as fit and speedy as their parents. This is a hidden problem, as in a world with an ever-larger population, there are many great young athletes. Their achievements are highly publicized, and we fail to recognize they are a tiny minority. In the same way, there are many millions who watch, discuss, and are fixated on football, cricket, rugby, or baseball, but their involvement in sport is totally as observers, not as active players.

Physical activity and performance are measurable, so these types of decline can be documented, whereas the psychological behaviour and changes in mental attitude are difficult to quantify. The quality of deep, long-lasting friendships suffers because of the reliance on rapid or superficial electronic contacts. Many will argue it is an effective route to meeting, dating, and marriage or partnership, but equally, it is obvious that there are a host of websites where the objective is to find a sexual partner for a very brief passing moment.

A further feature is that the Internet offers an easy access to porn, cybersex, and computer games and films where the objective is violence. Sociologists argue about how resilient young people are to exposure to such material, but there is no doubt that for many people, continuous viewing of these items means their moral standards are undermined. The consequence is a total lack of reality and sensitivity. They fail to understand how to treat others or recognize that warfare is not a screen game with images of bombs, explosions, and guns, but it involves real people, pain, and suffering. Teaching troops such insensitive attitudes via computer exposure is already used in military training, but reversal of that dehumanizing process for return to normal civilian life is ineffective. Therefore, this pattern implies that for the rest of the population, the exposure is a brutalization of attitudes, and a distinctly poor training for a civilized society.

To continue this list would seem repetitive, but there are also more subtle damaging effects. One such is that because electronic access to information is rapid and effective, many people I know (including mature intelligent ones), say they never trouble to remember anything anymore, as they can always recall it from their computer links. The weakness of this argument is that because they have not had to work to find and remember something, it is not ingrained in their memory, so it will not be valuable when they come to think about related topics (or want to impress their friends with erudition). The second aspect of relying on an Internet source is that, from many places outside of major cities, there is no access or Internet link. Finally, as I have sometimes discovered to my cost, websites that had some really useful information have been removed, or are no longer accessible, or they were updated, and the items that interested me have vanished. I have frequently used website science courses from other universities when preparing my own lectures (this is research, not plagiarism!), but later, when I decided to add more material, I have found the source had only been maintained for the duration of their course.

Isolating technologies for adults

For adults, electronic systems and technology have been called the 'digital divide'. Those who have it can benefit from communication and access, whereas those without are progressively more isolated. In parallel with the split between the electronic 'have' and 'have-nots' is invariably a gap in income and wealth. The dependence on technology just increases the gap between rich and poor; so it is a gap that is widening. Additionally, there are many people without physical access to electronic communication solely because of their geographic location. Therefore the split is not purely from education or wealth, although these are probably the most common reasons.

In terms of education, Internet access offers knowledge and reference material that was never previously available without travelling to specialist libraries. Learning foreign languages, or listening to discussions and music via replays of items that were broadcast or recorded elsewhere is a definite bonus. Internet access will help with gaining knowledge in topics as diverse as learning Latin, to understanding new techniques in plumbing and gas fitting. Purchasing goods at bargain prices, and finding opinions on the quality of those goods are further advantages. Such economies are precisely those that would be of most benefit to the poor, but are lost if they cannot afford or understand the technology.

They are similarly losing access to excellent government documents and information, which is progressively only available in detailed formats via the Internet. Electronic filing of income tax returns, vehicle registration and annual road tax, electronic banking, contacting gas or electricity suppliers, etc. can all become easier than sitting on a telephone waiting patiently for a person to answer a trivial question.

Medical and legal advice is an available benefit of the electronic databases. There is a further trend that service manuals and advice on home equipment are no longer included with the goods, but instead are accessible only on the Internet. For those with access this is excellent, as often the packages are much better prepared than earlier printed brief instruction sheets. Nevertheless, for those without the access, the situation is dire.

The spread of consumerism via the Internet inevitably means many shopping facilities have vanished from towns and villages. Not least are the closures of many banks. In parallel is a reliance on credit cards rather than tangible money. This can mean those with limited income will fail to realize that their electronic purchases are building up to become unsustainable debts. If this overspending leads to a block on their owning credit cards, then they have major difficulties, even if they have Internet access. The block is a standard problem for those who have been bankrupt. The 2014 US statistics suggest that around 70 per cent of those who could have a card do in fact own one, but about 7 per cent of people who want one are banned, because they have a bad track record for debts, or are bankrupt.

Seven per cent is again an example of a useful number for statisticians, but for most of us it carries no sense of reality. We should attempt to think in terms of numbers of people than it implies. For example, I could consider a city such as Birmingham (which is the second-largest UK city, with a population of a little over 1 million people). If I then guess that under half the population are adults who want credit cards, then 7 per cent of them means there are around 30,000 people who are banned from holding credit cards. This is an immense number of unfortunate people who are already deprived, and pressured into further difficulties because they are not part of the electronic money technology. Imagining the scale of 30,000 Birmingham people is easy, as this is the typical attendance at their St Andrews football stadium. A comparable US view would be a major league baseball stadium.

Job hunting

Once someone is looking for a job, then some computer skill and experience is extremely valuable, even if the work does not require it. The reason for this is that many companies are no longer advertising in newspapers. Instead they are using computer advertising or the equivalent within job centres. Such advertisements will then need an online application. This is a skill that people hope they will never need to practise too much, so it is difficult for them. Consequently, their attempts may poorly represent their worth and abilities; they will undersell themselves and remain unemployed. It therefore penalizes those without computer skills and training in making such applications. Further, there is a real risk that the application will be judged on the skill with words and self-presentation rather than on actual ability and background for the jobs. The problem is presumably worst for manual workers who may not need such computer and presentational skills at any other times.

A further limitation for the electronically nervous is that in some countries (particularly the USA), job interviews may now be conducted live on a computer link. This is a disaster for those who have no experience of such interviewing techniques. In my view, it is also a disaster for the employers, as they may not realize how poor they are at such interviewing; certainly the electronic interviews totally lack all the face-to-face information that is crucial when we meet people and plan to work with them. From my own experience, the divergence between the written comments and face-to-face interviews to decide the ability and suitability of job applicants has been surprisingly large.

The losers in all these situations are in the poorer-paid jobs as they involve people who are less likely to have learnt, or be able to use, the necessary application skills. Job turnover for this same group is also much faster. In more desk-bound, office, and shop jobs, the typical 20-to-35-year-old in the UK will change employer around every three years. Older people average out at rather more; say, five years. Indeed, unless an employee is making steady career progress through a big company or public service organization, then prospective employers respond less favourably to those who have stayed in the same job for more than five or six years.

This dependence on computational skills in advertising, applications, and interviewing has really only emerged for general use over the last decade. At the present time, it most obviously penalizes the lower-paid and workers without confidence with computers. Since the employers may not be maximizing their choices, the pattern of filling job vacancies may still evolve once the employers realize computer systems cannot guarantee to deliver the best employees.

One topic I briefly mentioned is that of electronics driving more advanced robotics. Here the progress is rapid and is not just confined to large car manufacturers, but is equally apparent in farming, where there can be totally automated feeding and milking of the cows, or satnav-controlled planting and harvesting of the fields without a driver on the machinery. All this is totally computer driven.

Those in this technology see the possibilities that for highly intelligent and skilled entrepreneurs, such robotics will be financially desirable and a source of great wealth. This is undoubtedly true. Their enthusiasm for the subject is matched by claims that it means boring, tedious, and dangerous jobs will no longer need humans, so people will have far more leisure time. Again this is totally true for those who have wealth. What these technologists fail to recognize is that entrepreneurs will never be more than, say, 10 per cent of the population; for the other 90 per cent, the skilled robotics means unemployment. The workers may be losing jobs that were boring, but at least they provided income and a sense of purpose. Robotics will definitely increase, but their presence will create social problems on a scale that no one seems to discuss. There is therefore a priority to do so. Mass unemployment as a result of robots may not be inevitable, but judging from the historic pattern of working practices in previous centuries, it is highly likely.

Access to fast communications—who really has it?

In the previous sections I have been emphasizing that, despite major benefits, there are also serious inequalities and destructive features of our dependence on electronic communications. Understanding this, and predicting where it is leading, is difficult, as fast electronic communications are a twenty-first century phenomenon. Isolation and poverty are interlinked with poor access to electronic technologies, and there are conflicts between the plus and minus aspects. Advantages of rapid communication are obvious, but systems as common as mobile phones and the Internet are not universally available, even in leading countries.

Marketing reports on the topic can be distinctly misleading as they always focus on the fastest systems available, and these are predominantly limited to major cities. Therefore, a claim that 90 per cent of the population has access to state-of-the-art high-speed sources is invariably distorted. The reality is that within large cities, 90 per cent of the socio-economic groups who are likely, and wealthy enough, to buy the fast products, may benefit. Further, fibre-optic links are definitely non-existent in large parts of both the suburbs and countryside.

Radio and satellite coverage is equally patchy and the data rates fall very noticeably. For satellite systems, there is a further complication that the signals can only travel at the speed of light and this introduces transmission delays. They may be acceptable for speech, but definitely it is not ideal for fast data rate communication.

If one is walking in many areas of the UK (not necessarily remote parts), one will find that access to satnav systems are non-existent, as the mobile phone networks do not (and cannot) give complete coverage. The option of a satellite phone may help, but they tend to be large and lack many of the normal phone functions.

The losers without broadband access are immediately obvious, and are very clearly apparent from the statistics in the USA. The country is large, so broadband services are inevitably inferior in rural areas. Further, there are a limited number of cable companies, and this means they have expensive access contracts. So those who are poor or live in remote areas are penalized.

Further, they lose out on the services over the range of items from video on demand, online medical access, knowledge, databases, and tuition and classroom facilities. For the poorer members of US society, the quality of affordable electronic communication will be slow and inferior to that which we assume is the norm. Unfortunately, this is a situation that will only get worse (e.g. if they are forced to use online job applications). Therefore, when it comes to education and opportunity, there is a major separation of between those with money and those without. Recent surveys claimed that only four in ten of US households with incomes below $25,000 per year had wired Internet access. By contrast, the numbers were well over 90 per cent for those households above $100,000 per year. The US numbers were closely linked to ethnicity, with white Americans being some 50 per cent more likely to have the linkages compared with African American or Hispanic households.

Overall, nearly 30 per cent of Americans do not have wired Internet access. In part this is availability; in part it is because the smartphone alternative is cheaper. Nevertheless, operational speeds are lower and downloads of data can take ten times as long. There may also be cash penalties for large file downloads by phone. The pricing is not competitive as each of the different companies often controls one region, and can freely raise the price of fast service.

This is definitely a weakness of the US system. As for other countries, from Japan to Sweden, there is government control of pricing, and high-speed operation is taken up by many more customers.

The digital divide is therefore increasing in the USA. Not least as US surveys reveal that the time spent watching videos or playing electronic games (often simultaneously) was nearly double for the poorer children with a lower educational family background. In the UK a similar pattern exists, but the total viewing hours are somewhat smaller. Unfortunately, the problem may be worsening, and the situation will not change until there is a more educated population, who demand better services than just TV and video. The difficulties are leading to a potential disaster both for the individuals and the nations.

Mobile phones are ubiquitous, but in some ways are more damaging to health than the larger screen and keyboards of computers. Yet more surveys in the USA say teenagers are sending or receiving around 100 text messages per day, and actually send more than 2,000 per month. UK patterns are similar, with a total of some 1.4 billion text messages per month. This activity has hidden health implications of bent spines, strained eyesight, and wrist and thumb strain (which afflicts roughly 40 per cent of keen text users). Other phenomena include a rapid rise in the number of replacement thumb joint surgery (linked to texting) and text addiction, with psychological evidence for insecurity, depression, and low self-esteem. Such features are reported from all the major countries across the globe.

Electronic access for older people

At the upper end of the age spectrum, there are the same problems of access caused by financial limitations, but these are compounded by the fact that many older people have grown up without ever having used computers. Therefore they are intimidated by them and lack the confidence to learn. They may additionally be concerned about the costs, and in many cases are very susceptible to electronic crime and failure to distinguish between genuine and spam-type activities of people who contact them. They may be equally susceptible to exploitation by people that they meet, but the written words of an email can seem more deceptive, and there is no visual contact to sense if the person is acting honestly or not. Hence caution in using electronic communication by the elderly is not unreasonable. Indeed, it is essential.

It is misleading to assume that the lack of Internet access is confined to the elderly. Data from the UK in 2014 suggest that one in five adults have never used the Internet. One in four do not have a computer at home, and, whilst for those under 25, Internet usage is well over 90 per cent, for those over 65 the number drops to around 40 per cent. The first guess would be that successive generations will be more computer and electronic literate, so the 40 per cent number will increase. I assume this is partly true, but the technologies are advancing faster than people learn and age, so it is not impossible that we are heading for even greater isolation of the poor and elderly driven by the new developments.

A second difficulty is that people are living to ages where their mental faculties are impaired, so computer usage becomes far too challenging for many. A sobering thought is that within both the UK and the USA, the number of physically impaired people, just with the mental distress of Alzheimer's, is some 5 million in each country. For the UK, that is nearly one in ten people and something that impacts virtually every single family. With extended life expectancy, the net effects of difficulties and care are rising.

Inflation and self-isolation

One underlying difficulty for older people, especially those who are retired and living on a pension, is that their mind set is geared to the income and living costs of when they were younger. They therefore see items, both tangible goods and intangible services, and products such as software, as extremely expensive. For a young working person, the same purchases will appear reasonably priced. This is inevitable as, for someone who is now retired, inflation has totally distorted their scale of values. For example, for an adult who is now 70, but still remembers costs and income in terms of initial teenage salaries, the cost of a pint of beer or a loaf of bread will have increased fifty-fold. Average salaries may well have matched this, but that is not the way we see prices, as they are coloured by our memories of when we were first experiencing them. House prices are of course far worse, as they have risen much more steeply. A cautious pattern of spending from older people is inevitable, as they realize that their finances are unlikely to improve, even if they are currently adequate.

This perspective is a disincentive to staying at the forefront of computer and Internet access. For those who only have a state pension, one software package could equal several weeks' income. For this category of people, there is a further hazard being proposed in which instead of buying software, it is on a contract that requires continuous monthly payments (even if it is only for very occasional usage). For the economic health of the country, this is a poor move, and there may need to be legislation to block such products, especially if it also implies that documents, letters, and photos that have been processed with such rented software can no longer be accessed.

I am concerned by this direction in thinking; for large companies, rented software may be desirable, and if this is a major market, then the software and service providers will see rental as an effective route to income generation, and not worry about the occasional user (particularly if they are 'only' elderly or poor).

Physical problems of computers and smartphones

Motivation for the elderly to use computers is not just reduced by reluctance to learn new skills, but also by the principle 'I have managed without it, so I do not need it'. The elderly are frequently unaware, or unwilling, to make online purchases as they do not appreciate the potential savings or cannot see the product in front of them.

Unfortunately, the efforts of many organizations to encourage older people to enter into the computer age is a mixed blessing, as for many it will initially be worthwhile, but eventually they will reach the point that they are extremely vulnerable to cybercrime and intimidation, precisely because they do not respond adequately to the challenges and threats associated with allowing electronic access into their lives.

Apparently the main computer and phone difficulty for alert elderly people is visibility, as they cannot focus for long at the distance of the computer screen, or have permanently blurred vision. Computers are designed by the young, for the young; visibility and manual dexterity are assumed to be that of young people. By contrast, a comfortable type size for the elderly is at least 14-point typeface, and both key spacing and button sizes needs to be larger than on a standard keyboard. An increase of around 50 per cent is helpful, but few people ever realize this. There are also more critical needs on the choice of colours on the screen. Our colour vision alters with age (not just failure of sensing the blue end of the spectrum), but also in discriminating between different shades.

I am still using the term 'elderly' very loosely, as in the UK we probably think this applies to anyone who is 20 years older than us, and certainly anyone who is retired (say, more than 65). Nevertheless, as part of ageing, physical problems develop, even if the mental state is excellent. Typically, hand control deteriorates, so touchscreens become tricky to use, and of course for those with large hands, or a lifetime of manual work and unfamiliarity with typing or computers, a lack of dexterity will isolate many sections of the population. These are all reasons there should be large keyboards for both computers and phones.

Age and mobile phones

Modern mobile phones have keypads and screens where the print size and key contacts are both too small and too sensitive for the elderly. The problems are both visual and mechanical. Fading eyesight and stiff fingers are just normal consequences of ageing. Tremors from diseases or age exacerbate such problems. Cataracts and macular degeneration are also common and afflict roughly half of those over 70. Electronic communication for this substantial number of people is therefore impaired. Fixed telephones exist with supersize keypads and a few preprogrammed, one-touch numbers for emergencies or special contacts. Unfortunately, there are far fewer examples of even fairly large keypad portable phones available. All those I have seen advertised appear to be very basic phones and have none of the application benefits of the 'young' person's mobile phone.

Mobile phone manufacturers have barely realized there is a major market opportunity for sales to elderly people. When I came to look at samples, I found that often the numbers were quite large (excellent), but typically the lettering needed for texting had not been increased, or in some cases seemed to be even smaller. This may have been clear to a young phone designer, but it totally misses the target user. Clearly, such companies need to employ elderly consultants and address these issues.

Rarely do larger-style phones have many of the facilities of smartphones, on the grounds that too many applications will be confusing. The users therefore lack the option to limit which displays and facilities are routinely available.

One very modest example of the need for a big keyboard plus a big-screen smartphone exists in my city. The parking meters are being replaced by a smartphone app to pay for parking. Hence I now know of a number of people who feel they are unable to drive into the city as they do not have, do not need, or cannot afford or operate the requisite smartphones.

One solution to shaky-hand control is to bypass it with voice recognition, both for commands and for writing. Voice recognition software has improved over the last few decades, but certainly there were many initial problems. Not least was that the software was designed by young men (often with Silicon Valley Californian accents), and it then had a high error rate for women and children with higher-pitched voices, and older people, or those with other accents. The latter problem also hits those speaking local dialects, or immigrants using a foreign language.

The overall message that emerges from the communications industry from many of these examples is that once you are retired, unemployed, or have a difficulty handling small phones, then the rest of the electronic world has forgotten you. In a society with an increasing percentage of elderly people, this is unacceptable. It is also a very poor business strategy, as there is a large market opportunity to sell products tailored to the very considerable number of those in the older generation. The only consolation is that the currently young programmers will also age into this group and may find themselves equally isolated.

To a young person, anyone over 50 is deemed to be old, and by 70 it seems incredible they still exist (I remember thinking much the same when I was 20). The reality is that currently there are some 22 million people in the UK over 50, with 14.5 million of them over 65, falling to 2.5 million over 75. The statistics for those who reach 70 in good health is actually quite encouraging, as roughly half of healthy 70-year-olds will reach 90. Centenarians are still a small minority, but in percentage terms they have increased more than tenfold over the last 50 years.

The point I am making is that young designers of electronics must focus not just on their own age group, but also see the needs of the market of more than 20 million in the retirement category. To put this in perspective, 20 million is four times the population of Scotland, or somewhat more than of all of those in the Greater London area. The phone sales marketing and peer group pressures may well be aimed at the under 15s as they are a sizeable and impressionable group who will buy the latest phone release, but in sheer numbers the over 65s significantly outnumber the young. This may not be image generated by magazines and the other media, but in sales opportunities, the more mature should be a prime area to offer electronic devices that are designed especially for them. Further, they should be at a reasonable price, as many big-button phones come with a bigger price tag than the equivalent small versions.

Predictive text

Much effort has gone into information technology; a fairly obvious way to offer help to those with unsteady hands is to use predictive text. Generally this is easier when using a keyboard, as it offers feel and feedback that is often lacking on a touchscreen. Tactile feedback is essential for people with poor manual control. Once again, ageism exists, as the predictive text is designed for the young, by the young. The predictive text on most mobiles is for a language and style that is incompatible with the more mature. If I use it, I realize I am not in the young generation. Predictive text in English needs to be fashioned and marketed to come in alternative predictive patterns that match the styles of different age groups and social patterns. The grammar, vocabulary, and slang of each age group are different.

This is certainly not a difficult technological challenge, as far more complex predictive versions already exist. For example, as briefly mentioned, predictive text in China uses the 26-letter, Latin-based alphabet to write in Pinyin (to make a sound variant of the Mandarin words). The predictive text cuts in after one or two letters to offer on the screen the most likely Chinese word character equivalents from an abbreviated set of the 5,000 main Chinese characters. (More Pinyin is needed for special characters.) The intended correct character is selected, and overall it offers a very fast text-writing route into the Mandarin characters. Indeed, it is generally faster than normal writing. On the receiving mobile, the text message appears in Mandarin. However, once again there is a severe problem with such character recognition on a tiny screen if the viewer is elderly or has poor eyesight.

Age-related changes to sound and light

Physical dexterity is not the only negative change with ageing. There are also effects on our voice, sight, and hearing. Typically our sensitivity to sound falls by 1,000 times (30 dB) between the ages of 20 to 70. This is not just a volume loss, as there is a much steeper fall-off at high frequencies, so sounds (and music) change in character. Many older people will fail to hear birdsong and noises that are apparently loud for the young. This is a standard ageing effect, but for the modern generations, it is exacerbated by conscious exposure to high sound levels. There is therefore an unfilled market niche as the sound electronics from mobile phones, radio, and TV need compensation for this change across the frequency range, but it is rarely available on standard equipment. Merely increasing the volume is inadequate.

As a musician, I recognize that my hearing is similarly showing signs of ageing. It is declining both in terms of loudness and high frequencies, so music has changed in character over the years, and for my CDs I now use a higher volume setting. At one time my responses were better: I clearly remember reading a book when I heard the sound of a moderate-size spider walking across the wooden floor 10 feet away from me. I was then in my 30s, so actually past peak sensitivity. Modern 30-year-olds may never even have experienced such sound sensitivity, as high-intensity music via headphones or speakers, visits to discos and all the other technologically driven sound systems, as well as city traffic, etc. rapidly cause permanent hearing loss. In terms of sensitivity, many teenagers now have hearing that is inferior to the 70-year-olds of earlier rural generations. This technology-driven loss of an essential aspect of a crucial human sense is indeed one of the darkest sides of modern technologies.

One should be critical of hearing loss that is caused by human activity and technology, as it is not a new phenomenon. For example, it has long been known that those who worked in noisy Victorian factories could rapidly lose their hearing. Legislation has helped at the industrial level, but no legislation can stop our self-destruction from listening to excessively loud music played through headphones or speakers.

Technology, colour vision, and ageing

In the realm of colour, there are problems that appear in an ageing population that are overlooked by young engineers who design and market electronic equipment. Colour vision changes with age, as does our sensitivity to light levels. For normal, healthy, mature adults, there is a need for higher light intensity than was required as a child; there are more problems with scattered light and glare; a slower response to changing light intensity; considerably worse contrast sensitivity and colour discrimination; and reduced ability to judge distances or focus on small print.

Computer screens have considerable glare for many people, especially with scatter from traces of cataracts or older eyes. Technology to help can be very simple, and I will suggest three options: (i) change the background from bright white to some other shade (e.g. pale green); (ii) wear polarized glasses, as the computer screen light is strongly polarized the glasses will dim the background; (iii) view the screen via a coloured filter (e.g. try transparent tinted wrapping papers). Cutting the range of colours will minimize chromatic blurring in the viewing. Narrow colour filters can also be helpful for some people with conditions such as dyslexia.

One of the fascinating ways we detect changes in our field of view is to use our eye muscles to continuously flick slightly. If we have a fixed gaze, then objects in our initial field of view fade into obscurity within a short time! This need to concentrate on changes is a consequence of survival instincts. Older eye muscles are weaker and the flick speed drops, with a consequent loss of fine detail in changing scenes.

In terms of intensity, a 70-year-old will need three or four times the light level required by a 20-year-old. The changes vary across the spectrum. As with sound, the higher-frequency range decreases the most. For light, this means that our blue sensitivity falls off faster than the red. Roughly, our blue sensitivity is halved between 10 and 20, a further halving by 40, and yet again by 70. This is very unfortunate, as technological advances in semiconductor light sources are impressive, but in applications, the focus has often been on the techniques and new colours, not on the users. Because it is now possible to make blue light–emitting diodes, they are incorporated in a vast range of equipment, from bedside clocks to displays on TVs and radios.

In terms of viewing, the use of blue light is a remarkably inept choice, as for many (not just older people), text is blurred and difficult to read when set into black plastic housing. For older people, information from clocks and displays is often totally lost with blue LED displays. Green is ideal, as it is our most sensitive colour region. We have evolved to respond efficiently to green as it matches the peak colour signal from the light transmitted through the atmosphere from the sun.

The other technological problem on many CD and TV displays is that the controls are embedded in a black plastic background, so have minimal contrast (for any age user). Initially, I assumed this was meant to be a design feature, but now I realize that making plastics can result in uncontrollable amounts of black flecks, so there is a higher failure rate when using, say, white background displays. The obvious mass-production solution is to use black plastic—then the specks do not show. So, overall, the display quality has been undermined by the production technology.

I will make two final comments on how marketing and displays fail to address all the population. The first is that some 10 per cent of Caucasian males are technically colour blind; the second is that perhaps half of older people have some level of cataracts. Colour blindness is actually a misnomer, as it is assessed by pattern recognition with, say, coloured numbers or pictures embedded in a background. For a 'standard' eye response, the colour patterns may show a picture or a number such as an 8. For those with a different sensitivity to colour across the spectrum, the pattern recognition will be different; they may fail to see an image or perhaps a 3 instead of an 8.

I am particularly aware of this, as I fail the standard test, but do so because I have a particularly good red response that extends further than the normal. Hence my colour sensitivity is absolutely fine for matching colours and for enjoying colour images. I take pleasure in seeing colour. This may well be the case with many others who are labelled colour blind, so they should not feel deprived. Indeed, we may justifiably feel superior as we have a greater spectral range. For example, when driving in misty conditions, the extended red response offers better clarity, as red light is scattered less than the shorter wavelengths of the blue and the green.

There are some people who genuinely lack the cone structures in the retina that respond to colour, so their world is in black and white. Whilst they lack the pleasure of seeing many colours, they invariably have compensation in that they have more rod-type detectors (which give the black-white contrast). Since these can be 100 times more sensitive than the colour detectors, their overall light sensitivity may well be far superior to the rest of the population.

Cataracts are common. They not only cause image blurring, but strongly absorb light at the blue end of the spectrum, so as the cataracts develop, images progressively move into being more yellow and then red. For anyone who wishes to see how the consequences develop, I recommend looking at websites that display images of the paintings of Monet. He had a small pond and bridge in his garden that he painted over the years. His pictures offer a remarkably accurate description of the development of his cataracts. The fine detail fades into very broad brush strokes and the spectrum progressively moves to the red. Eventually he had the cataracts removed and was horrified by his paintings, but medically they are a brilliant record of cataract problems.

Such subtlety in colour presentation and the corrections needed for it do not yet seem to be appreciated by those generating modern displays, although there is a clear market (unfortunately, many of the displays are not 'clear').

Will designers adjust their electronics for the elderly?

The realities of ageing mean items that are totally compatible for young people (the age group of the designers) may be inferior or inadequate for the elderly. This needs realization and a change in concept and motivation from the designers. It has not yet happened, but I am hopeful, not for altruistic reasons, but because of the vast numbers of older people in the advanced nations of the world, and the strength of the 'grey' pound, dollar, and euro. There is a large, rich market that is not being properly addressed.

Advanced medical centres

Technological progress is not just in electronics; unfortunately, its tendency to isolate vulnerable groups of society is equally apparent in other areas. I will take just one more example that particularly relates to the poor or elderly. Medicine is a major consumer of resources, and there is a trend to place expertise and expensive equipment in a limited number of sites. The logic and economics of doing this are quite obvious, as efficiency of using skilled people is improved. However, the implication is that the patients will be sufficiently fit and mobile to travel to these centres of 'local' excellence. Almost by definition, their patients are the people who cannot manage this. So the overall effect is that the technological advances have isolated those in need from the very sources of medical care that they require.

Whilst discussing such problems with a local care organization for the elderly, I was told that they are overwhelmed by such difficulties.

The local hospital, which is just a few miles away, has now moved many specialist units to a dedicated centre some 25 miles from the town. It is no longer directly accessible by public transport (even assuming the patients are fit enough to use it). For the new location, the triple bus journey takes more than two hours each way. Further, the centre makes appointments primarily in the morning from 8 a.m. onwards, which by public transport is an impossible journey, as the two regions are only linked by buses with poorly interlaced timetables. Therefore, the health group must fund taxis or private car drivers to transport the patients each way. The care group said this is beyond their budget, or that of most of the patients. The overall consequence is that a large percentage of the people are unable to attend treatment.

This is a pattern that is played out over the country for many elderly or poorer members of the community, or those who do not live in major cities. The failing is exacerbated by a blinkered view of the hospital organizers who tend to be younger, more affluent, with cars, or living within the major cities where public transport for the mobile is available.

It is also a problem that is increasing. In other examples that I have heard of locally, the overcrowded health services have been redirecting patients to centres of excellence that are some 90 miles and three hours distant. For the pensioners involved, the travel costs and difficulties of making such journeys have meant that many cannot attend for treatment.

Can we improve?

I have focussed on examples where, at each stage in life, technological progress has had negative side effects for different age groups. In all cases, the poor, weak, and elderly are penalized more than the rest. Whilst it is certainly happening—and indeed this has always been the pattern throughout human civilization—it is not yet clear if it is much worse than in the past. Speed of change has increased; so has awareness of the difficulties. Therefore, despite the apparent current difficulties in the reality of life, and the divides between rich and poor, healthy and ill, young and elderly, I am optimistic that we will continue to try to redress the problems. My confidence is that, in part, there is an untapped market for goods designed for the elderly.

13

Consumerism and Obsolescence

Obsolescence and marketing

In the non-human animal world, the main driving forces are food and sex. Sexual drive is the force behind fights between males for females. Even our image of the little Christmas card robins is misleading, as some 10 per cent of male robins are killers in fights over territory and females. We, as humans, have precisely these same instincts, but the thin veneer of civilization has added another layer to our problems—wealth and possessions. Technology therefore intrudes into our lives: as good technologies offer better and more desirable goods, we want them. This means that to purchase new toys we need money. According to one song, money is the root of all evil. Perhaps this is not totally true, but it is a high priority for most people, both rich and poor.

Social status and image

Affluent nations survive by skilful marketing and consumerism, but an undesirable feature of this is to have considerable obsolescence, together with the unfortunate social aspects of isolation (as discussed in the previous chapter). If we want to head towards a better world with a long-term future, then one target should be to minimize waste and obsolescence. This may seem unfashionable, as politicians and media all offer ideas and advice on how to increase our income, but very few seriously consider or propose either how, or why, we could economize. Politically, the reason is obvious, as it has little voting appeal, and similarly the general media and commercial TV need advertising money. I suspect I am a social misfit because I am unexcited by most advertising of new technology, but instead am concerned about the future.

Acceptable or commercially driven obsolescence

There are several types of obsolescence. The first is for products that have been superseded by a better item, or an item that removes problems associated with the original. Such obsolescence is fine. No one would now consider building an ice-house to act as a refrigerator, nor would we actively choose to buy a 50-year-old car for daily usage, unless we were able to maintain it and took pleasure from the work. In terms of hardware and products, whether cars, clothing, or washing machines, the goods will age, wear out, decay, and need replacement. This is inevitable, and we also age and decay or become ill. Indeed, obsolescence can be driven by genuine progress, and in the case of electronic items, we may well accept that the improvements justify throwing away older equipment.

Equally, if we have confidence and a strong character, then we can choose to resist, or follow, the fashion changes dictated by the style gurus. So fashion is also in the acceptable category of obsolescence, as we may choose.

Intentional obsolescence that is forced upon us varies from irritating to unacceptable. As a simple example, we can buy a floor mop in a supermarket, and discover a short time later that the replacement mop heads no longer exist, as the market has switched suppliers. The truly unacceptable examples are when we no longer have freedom of choice, and we are forced to throw away working items that were adequate for our needs.

If one looks at a technology where we, the consumers, are possibly ignorant of the details of the way a system works, at the industrial level the same situation is likely to be occurring, but hidden. For example, all our communications via optical fibres are forcing the cable companies to have signal capacity that is doubling on timescales of a year at most. Existing technologies can rarely be simply upgraded to cope with such changes, so the only solution is to invent new techniques. This takes time, considerable effort, and investment, so it is essential to recoup the development costs before they too have to be replaced. Speed is essential, as any delays can allow entry into the market by a competitor.

The only pragmatic and realistic solution is to ignore the earlier systems and go for new ones. This is equally a route to obsolescence, not just of the optical fibre technology, but also of many areas of the underlying science. This ongoing desperation at the development level will not be obvious to the general public. Nevertheless, failure to deliver will cause Internet traffic jams and chaos. So far this overload has mostly been avoided, but inevitably such events will occur. We will then need to totally review and reconsider what is essential Internet traffic. Perhaps a bonus to users is that we might then take the steps needed to block spam and unsolicited advertising, as it is estimated to be more than half of the email usage. Perhaps higher fee structures would seriously hit the spam and unsolicited junk mail, but still provide income to the cable companies.

Rather than only criticize, we should be impressed by the ongoing technological advances in long-range communication. Historically, with the visual signals sent by flags or heliographs, each pattern was limited to about one a second. So spelling a sentence with 60 letters could take a minute. This was the forefront of technology less than two centuries ago. The electrical circuits of the Morse code telegraph, in the mid-nineteenth century, raised it towards 200 characters per minute. The switch to valve electronics for radio and TV in the twentieth century represented a further improvement, by a factor of around a million times. Semiconductor electronics, together with modern optical fibre communications, now transmit as many as a hundred channels per fibre, each with data rates above a hundred million per second. So within 200 years, communication efficiency has risen more than a million million times. The progress is incredibly impressive, but achieved by totally discarding earlier methods and inventing new ones (i.e. controlled and conscious obsolescence). Unfortunately, we, the public, have responded by greater usage, and our demands are rising faster than the signal capacity. The potential for traffic jams or signal collapse are therefore increasingly likely in the moderately near future.

There is a similar problem with mobile phone signal capacity, especially because of the millions of photographs that are transmitted each day. Since their life expectancy is generally measured in minutes, it is not an efficient use of the technology. One may make similar comments about many blogs and other social media sites that are viewed by millions.

The communication systems are further stretched by surveillance of our transmissions, such as the government 'security' scanning of emails and Internet traffic for keywords related to terrorism or criminality. They are moderately effective, so in practice they are justifiable, even if we dislike the principle. By contrast, electronic eavesdropping to make profiles of our interests and lifestyles is unacceptably intrusive. In this case, the profile data are used to automatically provide related commercial advertising. However, both types of surveillance application add considerably to Internet activity. Even more annoying is that such intrusion is currently not just from our native governments, but is operated by other nations on both our internal and international emails, etc.

Less obvious obsolescence is caused by the quality of the manufactured product, especially of expensive items such as cars and double (or triple) glazing. Our only obvious clue on the build quality is the warranty. I interpret a short warranty as implying the construction is poor, whereas a long warranty offers hope of better construction. More caution may be needed if the warranty only applies to the purchaser and the product is typically sold on (e.g. if the house owner moves) during the warranty period. Warranties are a sensible guide, but not infallible, as there have been examples of highly popular car models that were sold at very low prices because costs had been cut in areas such as anti-rusting. For new cars, this is not immediately obvious, but with older models the patterns of value collapse reveal when this has happened.

Very rarely do makers ever say that they have built in a way of causing decay and the need for replacement. An exceptionally honest example was relayed to me from a friend of mine with a canal boat. He bought some new rope fenders that looked extremely well made and asked how the craftsman stayed in business. The maker admitted that there would be no need for replacements unless he encouraged them to rot, so he included some lime in the middle when he was making them!

Replacements before obsolescence

Manufacturer-driven obsolescence is forced on many of us by the markets of computer operating systems and software. In part, they may reflect improvements in the technology, but for many, the extra features available in replacement packages are non-essential. I carefully chose the word 'may', as the improvements being offered by the new software are often unwanted and, I suspect, for a vast number of us, the software packages contain far more gimmicks and processing power than we have ever used, or we may not even recognize that it exists within the package version we are running.

The only clear effect is that new software seems less efficient, as it consumes a greater chunk of the computer memory, requires frequent updates (maybe to fix errors that have been discovered), and means some earlier programmes are no longer functional. Quite often it will cause the entire computer to be inadequate, and we need a replacement as well as the new operating system. In such examples with mobile phones and software, the 'updating' is trying to force us to buy new, on a timescale of one or two years.

The really irritating features of 'upgrades', format changes, and new access devices is that our documentation that was stored on earlier formats and read with earlier software may suddenly become inaccessible. No matter how much the software companies pontificate that we should update everything to the new format, this is unrealistic. So, whereas at one time all our business letters, bank details, and family records could be securely kept on paper in a cupboard for the rare times that we needed access (e.g. for probate), all these data are now irretrievably lost. Worse is that the current ones are accessible to skilled hackers. Electronically stored photos will also be lost when software and formats change. The concern and examples of cybercrime that I cited earlier should equally make us concerned about long-term security of any records that we store only electronically.

I admit that some updates are essential, but a recent discovery of one major operating system revealed that it had an inherent weakness against illegal Internet access. This feature had existed for 20 years before it was spotted. Such revelations do not inspire confidence. External access to our private lives, computers, data, and references is clearly an increasing problem, and numerous international examples of major companies and government agencies being security breached merely emphasize that for the general user we are very vulnerable. This is not unfounded paranoia (e.g. I have already cited the example of the Kremlin spending 50,000 euros on electric typewriters, precisely to maintain security). We should not forget that banking details in handwritten ledgers have survived for centuries, as have historic documents. Maybe there is a new market for ledgers and record journals.

Armaments and warfare

Whilst economy-driven obsolescence is both incredibly wasteful of resources and costly, it is now firmly entrenched across the world.

Economic arguments drive it forward on the grounds that we need it to continuously fuel and expand sales and production. The fact that it is also an unsustainable approach is ignored at all levels of society. Our demand for novelty and instant satisfaction has blinkered us to our consideration for future generations.

If greed and consumerism were our only faults, then humanity might have developed quite differently. Unfortunately, just like the pretty little robins, we are totally uninhibited in killing our fellow creatures. These are not just other animals for food (perfectly acceptable if we were lions or polar bears and had no alternative), but also fellow humans. Avarice is driving us to take over their land and possessions, and force them to adopt our own views of how the world should be organized. This includes slavery and imposing ideological or religious lifestyles others, plus destroying their literature, culture, and language.

We have achieved dominance by killing, warfare, and genocide, which have been made possible by a range of advances in different sciences. A modest estimate must therefore be that improved technologies have killed or subjugated many billions of people over the course of our advancing 'civilization'. Once again, remember that history is written by the winners, so their cruelty and slaughter will rarely be documented.

For example, when learning Latin at school, children were often introduced to the simple books in Latin written by Julius Caesar. His *Gallic Wars* related how he overcame the Celts in France. He sometimes praises their military skills and bravery, but nowhere does he mention that prior to his campaigns there were around 3 million Celts living in the region we now call France. By the end of the war, his troops had killed 1 million, and taken a further million into slavery. The residual million had lost their culture, religions, and languages, and it was the end of the importance of Celts in mainland Europe. Caesar is presented as a great successful general, never as a practitioner of genocide.

Far from such military and political activity having reached an end point, both our efficiency at killing and devising ever more destructive weaponry are improving. Viewed in terms of technology, these are major advances. Indeed, we are so successful that it is not unreasonable to predict that there will be total destruction of humanity. Our current military technologies could already achieve this, and with the continuing vast investments in new weaponry, there is the danger that we may use these armaments in order to justify their existence.

To praise and catalogue our ingenuity and scientific advances in the realm of killing fellow humans is very easy, but it should be a total condemnation of why many technologies have evolved. Bows and arrows and flint knives for hunting were initial examples, but the early metallurgy of bronze and iron survive in archaeological examples of swords and armour. Therefore, we were already motivated to improve the metal skills for destructive reasons at a time when there was minimal pressure on us in terms of overcrowding. Similarly, the progress in steel making, as for Samurai swords, was highly skilled, but it was not intended or funded for better ploughs.

The pattern has continued on to explosives, machine guns, submarines, land and sea mines, and bombs. The state-of-the-art destructive power can now be delivered by remotely controlled missiles, or drones, and in terms of spectacular power, a single atomic bomb (either fission or fusion versions) can destroy an entire city. So, whereas one arrow or a sword might kill one enemy at relatively close quarters, and require courage as there was a high personal risk, the modern bombs are sent from a safe shelter, and can destroy a million fellow humans. Our innovative skills have also advanced warfare to include chemical and biological weaponry, aircraft and missiles with high speeds and long range, and accurate targeting via inertial, satnav guidance and laser-aiming systems.

The list continues, but the message is clear. We have always been willing to put money and effort and human innovative intelligence into destructive technologies. My initial premise of a dark side to technology may actually be totally incorrect. I am deluding myself, in line with every other commentator. Indeed, one could argue that we have made most progress in developing better materials primarily because we initially intended to use them to kill our fellow humans. From this more cynical viewpoint, the positive benefits of technology are therefore fortuitous spin-offs that are secondary to our initial objectives. There are elements of the same reversal of perception in the progress of medicine and biology, where knowledge was gained by treating wounded soldiers or gladiators, or advances and funding of prosthetics and plastic surgery, which were only enabled as a result of warfare. In reality, I suspect we cannot separate the two opposing views, as human nature has an equal mixture of negative and positive attitudes to our fellow humans.

If we pursue this discussion beyond the confines of Earth and ask if there are civilizations on the other planetary systems across the universe, then we need to expand our list of considerations on whether or not we can detect them. The equation of Frank Drake, which tries to estimate if other civilizations could exist, includes numerous factors, such as the number of planets that exist, and if they are habitable. The equation also includes uncertainties, such as whether they could develop intelligent life and communicate with us. Timing is also difficult, as such life forms could develop and then become extinct, so we might only receive signals from a long-dead historic intelligence. The universe is vast and signals from distant planets could take many thousands of years to reach us.

Despite these huge uncertainties, there are active attempts to receive alien signals, and simultaneously we are broadcasting our presence outwards from Earth. All our TV and radio transmissions do this, as well as more focussed attempts. In the half-century of our search to find signals of such alien societies, no signals have yet been found. This may actually be very encouraging, as any society that has developed the requisite technologies may be as self-destructive and expansionist as humans. Therefore if they find Earth, then we could expect invasion, exploitation of our planetary resources, and no more humans. The model would be precisely the same as for our earlier human civilizations that expanded and acted by destroying the people and resources of other continents.

If the aliens are slightly more intelligent than us, then they may not have the failings of greed, and the need to have power and dominate others. The fact we receive no messages from them is therefore excellent news, as it may imply they are a peaceful, idealistic, Ruritanian society. In that case, their technology will be focussed on other topics than interplanetary communications, space travel, and exploitation.

Knowing that other populated worlds exist would be exciting and extremely humbling for humans, but contacts with them might imply total disaster for us. Nevertheless, advances in the technologies used by astronomers over the last 20 years have identified the presence of planetary objects around other stars. This search is in its infancy, but already, within this very brief period of time, some 20,000 planets have been observed orbiting relatively close neighbouring stars. If we can already see this number from near neighbours, then when we scale up the numbers to match not just our galaxy, but also the myriads of galaxies that we can see, the implication is that there are many millions of other planets that could support intelligent life forms.

This possibility of many alien civilizations is unlikely to be directly relevant to life on Earth. Nevertheless, knowing that life forms may exist (or have existed) on other planets should definitely cause us to reassess our own self-importance. The universe was not created specifically for humans, and therefore we need to consider how we preserve our tiny fragment of the universe for the other creatures and life forms that exist here. Part of this reappraisal will be to limit our destruction of resources and living creatures. This means controlling our own population growth and reducing the demands on limited resources, as part of the present destruction is driven by our commercial activities, such as deliberate obsolescence.

For those who have religious beliefs, there is equally a need to reassess the scale of their thinking. Creation needs to be seen in terms of the entire universe, not just of one tiny infinitesimal planet within it. Hence their view of a Creator needs to be similarly scaled upwards. Copernicus was criticized because he realized Earth is only a planet within our solar system (not the centre of the universe). This dramatically downgraded the importance of humans from the narrow mindset of his time. Taking the truly universal view is a further, many million-fold downgrade. Nevertheless, more intelligent religious thinking will actually welcome the realization of the overall scale, as the alternative belief is to say that the Creator is many billion times more impressive than humans have been able to consider. Quite contrary to the general view, scientific knowledge is not rejecting creation, but putting it in a more sensible, and far greater, perspective.

14

Rejection of Knowledge and Information

How eager are we to learn?

In our inflated image of human progress, it may seem obvious that we will be keen to learn new ideas and skills. Reality is different, and we often actively reject ideas as well as factual new information. Initially, this attitude seems incongruous. Why should we not wish to learn new facts or concepts, or be so blinkered that we cannot comprehend them? However, the problem is well known and very apparent to all teachers, not just in schools, where children may be there somewhat unwillingly, but also at higher levels, including colleges and university. Rather than assume this happens solely from lack of interest or concentration, it now appears that we may be genetically programmed to ignore novel ideas. (This may already be a test of my comment!) Learning can be quite selective; there is a mental barrier except for ideas that are close to those we already understand. There are also some subjects that have a special appeal for us, so our inhibitions are reduced. For the rest, it is hard work or rejection. With this knowledge, and a better understanding of human behaviour, I feel I should now mentally apologize to the many students who seemed to ignore my lectures.

Self-taught skills are gained by experience and experimentation; for example, by tasting new foods. Caution is therefore a wise strategy for babies and children who need to experiment with unknown foods, but in practice it is equally true for new ideas. Caution is also sensible for most things in life—picking friends, making investments, starting new activities, or believing in new ideas. However, there is the same pattern: when we already have some related experience, we are more amenable to experiment and then branch out from a secure position into new territory. This cautious approach has been termed confirmation bias. We also may display a disconnect between our thoughts and actions termed cognitive dissonance. Overall it means that ideas far from our experience are automatically initially rejected, no matter how good the supporting evidence, whereas ideas or facts close to those we believe to be true will certainly be favourably considered and probably accepted. This may seem a little odd, but hearing things we understand is pleasant—it releases dopamine and serotonin, etc. in the brain, which we enjoy. New ideas have the opposite effect. Our liking for pleasurable sensation can equally colour many other decisions in subtle ways. Even driving errors and plane crashes have been linked to failure to recognize and make difficult decisions.

Rejection of new ideas need not be permanent, as given time we can gradually edge into accepting innovative concepts. Confidence in them is noticeably improved if we trust the presenter and source of the idea, especially if it is someone whom we believe is an expert in the topic. Parents, teachers, clergy, and TV stars may fit this 'expert' role if we have been conditioned to believe them, but, in general, novelty is low on our list of priorities, and we prefer existing concepts. Philosophy and religion are invariably based on opinion, so rejection of new input may not be surprising, but for scientific examples with hard and reproducible evidence, our rejection is less rational. Nevertheless, totally new scientific concepts are nearly always first rejected, and then they become slowly accepted over the next 20 years (i.e. roughly a new generation). Once in fashion, the novelty moves from way-out to become mainstream (even if it is wrong!).

In social terms, changes in attitudes can be very much slower than a generation. For example, in the UK it took nearly a century to legislate against slavery, and far longer to offer equality in other social terms. In many cases we still appear to have made no progress at all.

The bias to accept knowledge from experts is a tried and tested route to progress and knowledge, but it has some serious flaws. Not all 'experts' are correct, and if they are wrong, it is difficult to question them (they think they are experts). I want to offer new ideas, but they may be blocked by your instinctive reactions. So I will cheat. I have already made some comments on medicine, so I will reiterate them as examples. Subconsciously, this is not new material to you, so this time you will be more likely to believe it, and happy brain chemicals will flow.

This is definitely an efficient approach and it is widely used in religious and political indoctrination via repetitive mantras (especially at loud volume with repetitive music). In a large group of people, everyone else is saying the same phrases, and it becomes difficult to disagree; after a while you are trapped and believe the teaching.

Before the middle of the twentieth century, our attitude towards doctors and other people in authority was subservient and unquestioning. Because doctors had specialized knowledge, many people assumed they were infallible, or at least should never question them. It is only in recent years, with easy access to Internet opinion and knowledge, that dissent and a range of conflicting medical opinions have become evident to the general public.

Many examples are highlighted by TV coverage, ranging from statins and hormone replacement therapy to the effect of salt on hypertension. We want to trust experts because each style of treatment can have value in some cases, but equally it may come with long-term side effects in large fractions of the population. Sometimes the side effects are worse than the original problem. We hope experts will help, but in reality they may draw completely opposed conclusions, even from the same statistical data. For us, this is totally confusing.

In medicine and the biological sciences, the scope for misunderstanding or incorrect models, predictions, and use of new drugs is considerable, as there are very many factors that need to be considered. Furthermore, with living creatures, whether human or other animals, we do not have ideal and reproducible experimental samples. So if we are looking for a particular effect and we find it in some studies, we automatically see it as evidence for our ideas. The fact that other evidence contradicts us is then merely an example of abnormality. There are also difficulties in interpretation from, for example, statistics that appear to link the onset of Alzheimer's behaviour in old age to people who took antidepressants when young. One view may be that the drugs precipitated the condition; another view may be that the two factors were already related in the patient. Statistics alone do not help in such situations.

Another example, which from our perspective is completely false, was that in the 1920s women were encouraged to smoke, as it was claimed they would not gain weight. Statistically this may have been true, but for totally the wrong reasons.

New scientific ideas in engineering, technology, or medicine, where we have minimal expertise, are basically beyond us in many cases. So our only option is to trust our chosen experts. Unfortunately, the gurus in one branch of science may be completely ill informed in other areas, so even this is not an ideal strategy. There is also a conditioning and training effect that an established leader in a field will find it extremely hard to accept new thoughts in the same area, particularly if they imply the expert has been wrong over a long period of time. Ego and conditioning are far stronger than factual logic. Long-established experts have difficulty distancing themselves from their background and seeing new perspectives. By contrast, those who are outsiders may have no inhibitions, so spot new ways forward. The dilemma for outsiders is that they will have great difficulty in convincing the people who consider themselves already skilled and expert in the field. Nevertheless, outsiders may still feel intimidated in challenging official experts (e.g. in the way we defer to the white lab coat syndrome).

Expert pronouncements are not infallible, and we should be extremely cautious in knowing whom to trust, even if they have Nobel Prizes or run multimillion-dollar companies. There is considerable amusement in finding comments from famous people who turned out to be totally wrong, and a search of websites will reveal many examples. Before being too critical, we need to readjust our thinking to the knowledge available at the time, plus the fact that intervening progress may have changed the relative importance of their comments and perspectives. Nineteenth- and early-twentieth-century errors seem obvious, but there is no guarantee that our current twenty-first century predictions are not equally missing key ideas. Here is a list of unfortunate early quotations and oversights.

1840s: Colladon and Babinet separately invented alternative ways of bending light around curved paths, either in glass or water, and their ideas were used for fountains and stage lighting. However, the concept of applications in endoscopy and communications was thought to be impossible. The views of industrial leaders had not changed much even by the 1960s, as optical fibres were seen only as a laboratory gimmick that could never displace radio communications.

1876: The telephone fared no better as it had 'too many shortcomings to be seriously considered as a means of communication. The device is inherently of no value to us' – Western Union.

1878/80: Electric lighting similarly struggled: 'When the Paris Exhibition closes, the electric light will close with it, and no more will be heard of it'; 'it is a conspicuous failure'.

1883: 'X-rays will prove to be a hoax' – Lord Kelvin (then president of the Royal Society).

1903 'The horse is here to stay, but the automobile is only a novelty—a fad' – comment from a bank against investing in the Ford Motor Company. Indeed, in 1903, with poor roads, this was not such bad advice. A real incentive for motor power came from the dark side of technology: weapons and tanks killed several million horses during the First World War, and their vulnerability for use in warfare was exposed.

1946 'Television won't last' – hopeful thought from 20th Century Fox.

1959 'The world potential market for copying machines is 5,000, at most' – IBM. For the machines then being produced, this was fair comment.

1977 'There is no reason for an individual to have a computer in his home' – DEC. This was not an unreasonable view, as at that time machines with significant processing power were immense and costly to operate.

Most such pronouncements are the result of ignorance and lack of foresight, but some errors are driven by wishful thinking or bias. I have seen bias or blinkered views many times when acting as a scientific consultant. Invariably, I am initially extremely ignorant in the field compared with those I am advising. It is therefore essential to ask questions very carefully on what is the difficulty that they want to overcome, how have they had proceeded in the past, and why those methods were favoured. The typical response is they have always done it that way, and no one ever questioned it. As an uninhibited outsider it is then possible to offer new ideas. In these consultancy roles, the in-house commercial experts are willing to listen, as they are paying, or have approached me for advice. If an unsolicited outsider had volunteered the same thoughts, the companies would invariably have ignored the ideas. I am told by wiser consultants that companies are more receptive if the consultant charges a higher fee.

This is not a new idea: increasing the price to raise the desirability of a product is a familiar example of successful marketing and it works in areas from perfumes and clothing to cars and holidays—it is not an unexpected phenomenon.

Rejection from distrust, religion, and culture

I have just mentioned the responses where we find it difficult to accept new ideas that at are far from our previous experience. To me this seems a reasonable attitude, but I am more surprised that we can strongly apply a total disbelief to ideas or information, even when we can see tangible evidence in front of us.

One of the unexpected responses of human behaviour is how strongly we can apply our prejudices and conditioning to totally reject and not believe in facts, not just reported to us, but ones where we can also examine the evidence. A very remarkable example, which I saw on television, was a programme discussing the value of recycling waste materials. The logic of this concept is obvious, as it can minimize exploitation of new resources, increase availability of raw materials, minimize processing costs, reduce the scale of landfill waste disposal, and, not least, when well managed, offer financial benefits to councils that do this efficiently.

The programme interviewed a group of articulate people who not only did not make any attempt at recycling of their waste materials, but emphatically stated their total disbelief that the local councils actually took any action, despite the official public statements. Distrust of some councils or central administration may be justified, but totally surprising was that having seen the recycling plant, they still felt it was solely intended to reduce the volume of landfill! Only after being presented with products made from the recycled material, and been informed of the scale of income that it generated for the council, did they manage to reassess their prejudices, and realize it was valuable on many counts, and a more sustainable way to handle the rubbish.

In principle, efficient, well-organized recycling should reduce the drain on natural resources and therefore is to be encouraged. For the public, a difficulty is that such schemes are variable, so no clear pattern emerges. The TV example additionally showed that the relevant council had failed to effectively tell its public how efficient they were and how profitable it was to their city, and to emphasize the more general benefits of the policy.

Excessive reliance on initial opinions

We are also blinkered if we believe we already understand a problem or an idea, because then we either overlook key information, which is in full sight, or reinterpret it or dismiss it so that it fits our preconceived views. Again this is not prejudice but inherent human nature. This unexpected and subtle reason for rejection of information comes from our skill and ability to make a rapid value judgement when we first meet a person or situation. It is a good strategy, as we rapidly separate out a possible friend or a threat. Unfortunately, once we have decided, we subconsciously look for subsequent evidence to reinforce our initial view.

If we gain more information that contradicts our first view, then we have several choices. The first option is to ignore the new information; the second is to try to reinterpret it so that it fits our preconceived viewpoint. In either case, we are deluding ourselves. The third possibility—that we were wrong—is very low on our agenda.

Inability to accept being wrong is classic human behaviour (particularly for those who are insecure), and it seems to run across the whole spectrum of our dealings with people and facts. In science, there are many examples of excellent scientists who have missed known key information that would have helped them. In social contexts, we make the same mistakes. A particularly serious, and a little-publicized, frequent example is that made by members of a jury in a court trial. In times of danger, we have needed the instant decision of 'fight or flight'. So we automatically and unwittingly make initial rapid assessments of the guilt or innocence of the accused on first sight. Jury members then tend to overlook evidence that contradicts their initial view. So look innocent when you enter the court! There is no second chance to make a first impression.

In studies of how doctors scan an X-ray image to look for cancers, it is possible to track their eye movements and see how diligently they view the entire picture. A very common situation is that, once they detect a suspicious region, this absorbs all their concentration. They will think that they are continuing to scan the total picture, but in reality their eyes only briefly flit over other parts of the image (which may include a second cancerous site). Having found one abnormal site, the usual behaviour is to focus on it and exclude any other information in the picture. This is not just the behaviour of inexperienced doctors but equally a failure of experts.

Further, even the best of multi-tasking brains has a very small limit to the number of tasks and bits of information that can be handled. So if there is a lot of complexity in the image being scanned, we cannot cope. This inherent human weakness (not just in medical images) is one of the justifications why there are benefits in using computer-based pattern recognition, as here the human traits are avoided and every element of the picture will be equally well scrutinized. Once the highlights have been identified in this way, the human expert will happily accept, and view, all the crucial features. Maybe this is because the attitude towards the information is less personalized, so they do not have inhibitions or attachments to a limited focus on one area.

Rejection is equally apparent when the new ideas and information clash with thoughts that have been ingrained from childhood by the local culture, place in society, or religion. Here one can find a host of highly contentious examples, but rather than consider an extreme example, I will just comment on one that was caused by scholarship in the seventeenth century. An Irishman, Bishop Ussher, used literature examples available to him to try to estimate the age of civilization. He succeeded in demonstrating that there were well-documented written records going back to at least 4000 BC in the Western world. (He did not have access to older writings from, say, China.) However, his age limit was then later presented as meaning it was linked to creation, and therefore was the age of the universe. The underlying problem is that we, as selfish humans, wish to believe we are the most important creature in the universe and that it was created specifically for us. Effectively, it is a 'spoilt child' syndrome.

Any evidence that demonstrates humans were merely a species that evolved from earlier life forms downgrades this self-importance, and therefore many people reject such possibilities.

Viewed with twenty-first-century data, we can now say that Earth has existed and evolved over billions of years, and the universe, as far as we can detect it, has expanded over some 13.8 billion (thousand million) years. These are scientific data that were not available to Ussher. It of course offers no comment on creation by a deity, except that instead of creation being specifically for our benefit, we now need a far more impressive event for the entire universe. From the spoilt-child view of humanity, we have sunk into being a very minor part of the total universe. Hence the instinctive reaction is to say scientific data must be false rather than trying to face up to, and comprehending, the unimaginably large scale of reality. In fact, this is not the end of the pattern, as astrophysicists try to think beyond the detectable universe to ask if there were forerunners, or hidden but parallel universes. For most of us, this is a challenge beyond our comprehension.

Information loss from an excess of data

If information is gathered at an extremely high rate or in very large quantities, it is virtually impossible to process and analyse it. A well-documented example is satellite images, which are employed by many nations to survey military movements in other countries or to check on agricultural output from different areas. When monitoring military sites and troop movements of another nation, the satellite data can be generating results from several thousand sites per day, and doing so in considerable detail. Modern systems have no difficulty in resolving a single truck or tank, and movements of such vehicles are informative in military terms. With thousands of images per day, the results need to be viewed by many hundreds of different people. Therefore, the normal practice is to discard much of the data to make it manageable. Of the data retained, each observer needs to decide what is important in their set of images and feed these data to a higher level of management. Eventually the top level needs to correlate and make some sense of these snippets of information. For surveillance of a major country, this is unrealistic, and I have read that quite commonly as much as 60 per cent of the information is never even viewed, and much of the rest is just stored for future reference. Decision making is then back to intuition, prejudice, direct information from ground sources, and good fortune. In parallel with this will be demands for greater funding and more viewing satellites, etc. (No, I am not being cynical, just accurate.)

Precisely the same types of excess of data are encountered in many other surveys or sets of measurements. For example, in the field of particle physics, people are seeking evidence of unusual reactions that are captured by equipment that generates stored data at phenomenal rates. Human analysis is totally impossible, so there is software written to hunt through the data to try to find and identify specific events. For familiar situations of the more common types of process, this works reasonably well. However, if the software has not included searches for an unexpected and novel event, then the analysis will not reveal it, and it will be missed and lost.

Whenever a search of a very large database is made, the success depends on the foresight of the writer of the analytical software. Computer search engines are not themselves capable of seeing original patterns. Rather than criticize such computer-based searches, the particle physics examples show precisely why they are needed. In the hunt for the particle termed the Higgs boson, high-energy protons were impacted together, and from the fragments of the proton collision, one of the by-products is potentially the Higgs boson. It is definitely not a normal fragmentation process: it occurs only about once in a hundred trillion events. To detect it enough times—to be confident it is not merely a flaw in the data or computations—means we need a large number of positive identification events. On this phenomenal scale of data processing, computer analysis is the only option.

An example of plate tectonics

The patterns of behaviour that I have just mentioned are quite general, but examples can offer the scale of the oversights. I will start with the problem of how we discover or accept new ideas. I will use an example of rejection because the idea came from an outsider to the topic area. This was the case with the idea of plate tectonics. The theory describes how entire continent-size blocks of the earth drift across the surface of the planet. The credit for the idea should have gone immediately to Alfred Wegener. He was a German meteorologist with a very wide perspective of scientific observations. He noted that the shape of South America roughly matched the outline of the coast of Africa (indeed most of us spotted this when we looked at maps at school), but in addition he looked at the fossils on these matching coastlines, plus their sequences of rock layers. He found close similarities in each case. Whenever the shapes fitted, then so did the fossils and the geology. For example, the lower strata of rock types might be granite, underlying slate, and sandstone, on both sides of the Atlantic. Wegener then made the mental jump to suggest that in some ancient historical time the two continents had been joined, but they had split and the two sides had drifted apart. He published his views in a book in 1912. Unfortunately, he was not a geologist (i.e. he was in the wrong field), and also because the idea was totally novel, his work was rejected or rubbished. Worse, his reputation was damaged, and he had to hunt for a new job. The publication of the work was also unfortunately timed, as Germany went to war in 1914.

A further problem was that Wegener was sensibly fitting the continental coastlines, which differ somewhat from the current seashore shape. His critics did not understand this, and in turn he did not understand their comments, because they were in English. He was not vindicated until the palaeomagnetism work of the Indian tectonic plate movements in the 1950s and the underwater submarine studies in the Atlantic in the 1960s, which detected the volcanic activity that is causing the separation of the African and South American continents.

Interestingly, the submarine studies were not concerned with geology—the data were collected as a result of military studies related to submarine warfare. The details of this mid-Atlantic ridge are highly detailed and have even added the timescale of the movements of the separation. Wegener's model is now fully confirmed and listed as standard textbook material. Poor Wegener never knew this, as he froze to death on an expedition.

As I have already mentioned, tectonic plate movements are extremely important as they are the driving force for many volcanoes and earthquakes. It is therefore surprising that we only started to understand and believe this in the last half-century.

Difficulties for Copernicus

Copernicus had a number of obstacles to overcome. The first was that he wanted to present an idea in which the sun was the central point around which the planets all move in roughly circular orbits (or at least elliptical ones). The selling point was that such paths were much simpler to understand than the complex epicyclical tracks one would plot if the sun and planets circled Earth. In fact, this was not a totally novel idea, as a Greek (Aristarchus) in the third century BC had considered a heliocentric set of motions. Unfortunately, Copernicus was a priest, and the Church believed that if the earth was not the centre of the universe, it undermined the central importance of mankind (and especially the Church) as the epitome of the universe. Consequently, Copernicus presented his ideas with some caveats several decades after he had considered them, and his book only appeared in the year of his death. He also dedicated it to the Pope as a wise political move.

The second problem in accepting his theory was a perfectly valid scientific one at the time. There were excellent astronomical measurements by Tycho Brahe, who realized that a heliocentric model was very successful for the planets orbiting the sun. However, the model predicted that because the stars appear to be fixed, they must be at immense distances from the solar system (i.e. much farther away than any of the planets). Also, it was unlikely that all stars were at the same distance.

Nevertheless, the sizes of the stellar images that we see are all very similar, and not much smaller than those of the planets. Effectively, we see star images that are far too large for really distant sources.

Brahe was correct in his comment and criticism. It was only two centuries later that our understanding of light had advanced and we realized that it has properties of waves. This means that our knowledge had improved to the level where we understood that for tiny images, the 'point size' that we see is set by the optics of telescopes and our eyes (not the distance to the original star). Whilst distant objects we see on the surface of the earth look smaller than similar ones that are close to us, if the object is extremely far away and the image size is minute, then we need to add in more complex physics. It is now well understood and involves the wave-like properties of the light.

Overall, this is a nice example of how we can reject ideas because we lack all the necessary information, rather than from prejudice.

Whom should we believe?

I have mentioned that we change our level of acceptance (or rejection) of ideas depending on the status of the person who is making them. My examples are deliberately science-based, as in these cases the input data can be measured, repeated, and quantified. This offers hope that we have made the correct decision. Nevertheless, Wegener lost out because he was in the wrong field and it was difficult for other people to test his theory. By contrast, Georg Ohm was a teacher who, in 1827, published results that, in today's language, link the electrical current (I) going through a piece of metal, the voltage (V) that drives it, and the resistance (R) of the object as ($V = IR$). Today, no one would question this in simple electrical circuitry, but in 1827 he was totally rejected as the German philosophy at the time considered experimental results unnecessary. Also, his ideas were in some conflict with mathematical work in France. Basically, the law was too simple and understandable, so he was considered unsuitable to be a professor.

Conversely, there is a danger that status outweighs critical comment, and in a few cases our reverence and admiration can block normal caution. Einstein is ranked as probably the greatest scientist of the twentieth century, and in his own research fields this is fully justified. However, he became interested in a suggestion by Hapgood that movement of Arctic ice might suddenly cause crustal movements of the earth, and they would do so on a rapid timescale. This idea by Hapgood seems to ignore, or be in conflict, with the tectonic plate theory of Wegener. Nevertheless, Einstein wrote a cautious supportive preface to the book in which Hapgood discussed his theory, and because of the eminent status of the writer of the preface, the theory temporarily became fashionable. Eventually it was discredited.

The general relativity work of Einstein is a fundamental part of modern cosmology and is unquestioned in terms of its relevance to the ongoing theories of the creation of the universe. This seems to be true, but rarely does anyone mention that the theory was developed many years before the bulk of the experimental observations were made that support modern cosmology. It may be sensible (or heretical) to ask whether modifications should be considered in the light of our current knowledge. Because of the scientific status of Einstein, any reappraisal seems unlikely, as few people would ever consider risking their reputations in such a discussion. However, it is inevitable that eventually reassessments or additions will be made.

We may imagine that the superheroes of science are the Nobel Prize winners, but unquestioned faith in the work of Nobel Prize winners is unwise if it is not part of their expertise. There is a very salutary example in the claims of Linus Pauling. Pauling was undoubtedly an excellent chemist. He is one of only two people to have received two unshared Nobel Prizes, and therefore doubly likely to be believed no matter what topic he discussed. He made extreme claims for the role of vitamin C in that he said it could prevent and/or cure the common cold; later he said that it was effective in cancer treatments. These claims helped sell many of his books and brought awards and the founding of an institute. The vitamin C industry that sprang from his ideas is estimated to be on the scale of hundreds of millions of dollars per year in the USA.

Unfortunately, modern reappraisals of the evidence he used for and against the efficacy of vitamin C suggest he was highly selective and prejudiced. Indeed, many later studies have unequivocally refuted his claims. Despite the facts, the industry thrives.

I am claiming that examples from science are potentially less contentious than matters of philosophy or politics, as in opinionated topics there are always conflicting views rather than secure data. However, one extreme end of the range of topics in science is mathematics. Advanced mathematics is, for most of us, something that requires hard work and often considerable intelligence. This has several consequences.

First, intuition is normally unreliable, and it is difficult to challenge a mathematical idea or proof without working through all the steps and considering alternatives. Second, we lose face if we admit we cannot understand it. Therefore for the majority (including professional scientists), the easy option is to assume a complex mathematical proof is correct. In this case, we are making the mistake of accepting an idea for the wrong reasons.

I recall a classic example of such an error from a lecture I heard by Paul Dirac (a Nobel-winning theoretical physicist) whilst he was addressing a very large scientific audience about his early work. He started by saying, 'I realize I am only talking to a few of you' (a comment that was both humorous and true), and then cited a famous equation that he had written at the end of the 1930s. It was then widely employed. He pointed out that despite this usage there was a mistake that no one else had noticed until he realized it himself some ten years later. His reputation was such that no one had thought to doubt or challenge his work.

The conclusion to draw is that sometimes we are overwhelmed and assume a new idea from a prestigious source must be correct, especially if we are embarrassed to admit that we could not fully understand it. A familiar situation is that the use of foreign words means we do not understand and take the easy option of not challenging what we are being told. Many people have claimed that doctors, lawyers, and clergy deliberately use Latin-based words precisely to inhibit discussion.

There are also cultural oddities involved in a teacher-student transmission of ideas; for example, the student knows the teacher is wrong but is unable to say so because of their relative positions. I have experienced this quite a few times as many of my students have been brighter than I am (and probably still are). When I made mistakes, some students were extremely reluctant to point them out. Eventually, I realized my errors and then asked, 'Why did you not tell me?' Depending on their background culture, the answers varied: because you are the teacher; you are older; or even, you are a man and I am a woman.

In terms of information loss, or distortion, all these examples emphasize how difficult it is to make progress. If our basic information and understanding are faulty, then we undermine our knowledge and block progress, or inhibit any subsequent strokes of inspiration and deep insights.

Information rejection from geographic isolation, xenophobia, religion, and prejudice

Examples of deliberate destruction of information associated with wars and religion are depressingly familiar, but in addition there are many more subtle examples where information and knowledge are dismissed because of xenophobia. The prejudices come in several layers. The first may simply be because the facts or ideas are in a foreign language, of from a country or historical period where we have few contacts or little understanding. Such ignorance has resulted in many examples of the same progress being made in countries or centuries without realizing they are reinventions. For example, the well-known theory of Pythagoras about the relative lengths of the sides of a right-angle triangle is now acknowledged to have been known several centuries earlier before the writings of Pythagoras. The earlier version was written by a pyramid builder in Thebes in Egypt.

Pythagoras had spent some time in study in Egypt, so it is unclear if he had any knowledge of the pyramid writing, or whether he totally reinvented the theorem, but certainly he must be credited with a clear proof and wide dissemination of the idea.

Other science examples exist. Western mathematicians assumed that no earlier civilizations were capable of deep mathematical progress, whereas in reality mathematics in China, Egypt, India, and Persia had often been far ahead of the West. Two classic examples cited by modern mathematicians is the fact that the concept of zero and a value for pi (π) had both been derived many centuries earlier than in the West.

Ignorance from lack of contact is common across all of our cultures, from literature to arts and music, as well as scientific knowledge. Music offers many examples of this isolation from distant work, and here it is definitely not a problem of language, as musical notation is global (at least for Western classical music). Nevertheless, if one listens to 'classical' music in the UK, there are a remarkably limited number of composers who provide the bulk of music that is broadcast, or even played in most concerts. This either implies that great composers are incredibly rare, or that we have forgotten many others of equal quality because of changes in fashion or other isolating factors. There is an element of filtering in that we frequently are presented with only the best-quality foreign composers, but are more tolerant and include lower-ranking local work, because of inherent nationalism. This is only part of the limitation. In many cases, music critics will idolize one composer or performer, and the effect overshadows great works from contemporaries or recordings by other artists. For example, if asked to name Finnish composers, most people will cite Sibelius, but be quite ignorant of his contemporaries.

There is a further restriction imposed by at least one classical music channel: they have a list of the top 100 or 200 tunes and then vote on these so that they are played more often. So the five-minute fragments are incessantly repeated and, not surprisingly, these are the only items that are voted into the top preference list. The commercial marketing has driven them into totally safe territory, with minor efforts to popularize the wealth of excellent music that exists beyond the top favourites.

The value of publicity for a particular composer or musical style should not be underrated. Bach is currently highly ranked, but within 50 years of his death his work was almost forgotten, except by those undertaking musical training. However, a century later, Mendelssohn, who was a fashionable composer at that time, generated a revival of interest in Bach's music that has survived to the present. Many of Bach's contemporaries were more highly rated during his lifetime but, without a champion, they have faded away in our collective memory. Public taste is fickle and easily manipulated, so we can be robbed of great and rewarding music. The same situation is undoubtedly apparent in art and literature as well. In each case, these are examples of information loss caused by human activities, not by physical loss of the materials on which they were written.

In parallel with such engineering of popular appeal, there is ignorance, because the compositions of less familiar composers have not escaped from their homeland. I am very keen on classical music, but am frequently amazed at how often I come across really enjoyable music that has been written by composers who are rarely mentioned or broadcast. In my local music library there are many CDs from the usual fashionable names, and a handful of items from these forgotten composers. Since the quality of their work and performance is often excellent, it is not obvious how this has happened. One pattern I have noted is that such composers often came from periods when their home countries were under the control of foreign invaders or political dictatorships. This presumably meant they could not travel to the outside world, and nobody was encouraged to visit.

One such example is the contrast between two Polish composers, Chopin and Dobrzynski, born in the first decade of the nineteenth century. They were contemporary and students of Elsner in Warsaw. Both were excellent, but Elsner rated Dobrzynski as the better composer. Chopin left Poland and toured the rest of Europe and very competently gained fame, prestige, and a long legacy of followers. Dobrzynski stayed in Poland, which was then plunged into political turmoil from an uprising in November 1830. He survived and wrote music, but because of the political isolation, his work only seems to be known and appreciated by fellow Poles. His pupils and colleagues have suffered a similar fate as viewed from the UK. He and the many in similar situations from countries across Europe have an unknown legacy for the rest of the world. For me, this is a side effect of information rejection and loss from both political events and nationalistic exclusion.

Ignorance of a wealth of musical composition exists even though no language is involved. In any area that involves language, from philosophy to science, ideas and information are far more likely to include works of value and interest that are unknown, not least because of the need for translation in terms of both language and context at the time of origin. Music is for pleasure, but as a scientist I want information related to my field of research and I do not mind where it comes from. For me, the country of origin and their politics are immaterial if the results and ideas are well presented. I am clearly unusual, as I have worked in several countries where this was not the attitude of major laboratories.

In one Western example, the library would not subscribe to journals from the Eastern bloc, as they claimed the 'foreigners' were inadequately funded, so could therefore not have original ideas and data. In other cases, I have seen political antipathy and refusal to read work from a specific region. Having also worked in the opposing country, I then noted precisely the same attitude in reverse. Such mutual xenophobia is particularly sad as it is in a scientific field totally dissociated from the religious and political differences between these nations.

Traces of prejudice and xenophobia are often apparent (even from nominally liberal people), and they appear in the 'impartial' world of science as easily as in politics and daily life. Scientists spend a great deal of time writing begging letters, although they are officially called grant proposals, and having been funded, they try to establish their reputation by publishing their results in highly rated journals. In both cases, appraisals, funding, or rejection are supposedly by unbiased experts in the relevant areas. Unfortunately, impartiality is often compromised either by the name, reputation, country, or institution of the submissions. Famous-name people from prestige institutions benefit from this, so have a self-fulfilling high acceptance rate. The problem is well understood; some journals have conducted experiments where the names of authors and institutions were altered for the items sent to referees. As expected, for precisely the same articles, acceptance was more immediate and with fewer critical comments for work labelled with highly rated contributors or famous institutions. More critical comments and rejection were given to the articles if they were attributed to unknown authors or institutions that were perceived to be of low status.

Funding and publication of science also seem to be easier in fields that are very popular, even if the work is mundane, whereas novel items draw far more criticism. Having understood the problem, it is possible to improve success rate by including prestigious names and institutions on grant proposals and publications. In my experience, this window dressing has been helpful.

Names are equally critical and quality is unfortunately assumed by the number of science citations, especially for those who are first author on the article. In some countries, the practice is to always write the authors in an alphabetical sequence, but for the rest of the world, who are ignorant of the practice, we misunderstand and credit the wrong people. Names starting with letters such as A, B, etc. also gain benefits in many job interviews where people are seen in alphabetical sequence.

In job interviews, and also in musical experience of comparing the quality of instruments, our judgements are distorted by knowing that we must make a decision from, say, a group of five. The music examples are often made with the player behind a screen and playing, for example, the same music on five different violins. Initially, we are cautious and think we must wait until we hear later ones; finally we become bored. So there are a disproportionate number of winners for a middle placing. This even happens if the same instrument is played more than once! Maybe less surprising is that if we are correctly told the maker of each instrument, then the famous maker has a higher chance of being ranked as the best. Deliberately misleading the audience by incorrectly naming the instruments will shift votes with the name, not the instrument.

Names that contain emotive words similarly distort our judgements. A *Scientific American* journal article discussed this in the field of conservation of endangered species. They cited contrasting enthusiasm for saving different creatures. The examples included the patriot or killer falcons; American or sheep-eating eagles; American or hairy-nosed otters. In each case, the former were supported around 50 per cent more than the animal with the less attractive or emotive name.

Equally obvious are the examples of immigrants using a new name that is more typical of the host country, either because it is easier to spell or pronounce it, or because it had too many overtones associated with it. A classic example at the time of the First World War is that in England, the royal family had the good German name of Saxe-Coburg-Gotha, which had come via Prince Albert. Queen Victoria was also of German origin from the House of Hanover. In terms of European monarchies, both indicated excellent pedigrees. However, since the war in 1914 was with Germany, this was not an ideal name for patriotic support, and by 1917 the family rebranded as members of the House of Windsor. Windsor is a very British town with an impressive castle, and close to Runnymede, where the Magna Carta was signed in 1215.

All such examples underline the fact that we do not make rational judgements and our prejudices exist without us realizing it.

News coverage

The facts, news, and opinion fed to us by the media are inevitably skewed or limited by those who influence or control the media and present it. Most of us have a strong parochial streak and are interested in very local events, especially if they include places and people that we know. This selective news coverage passively or actively shapes our opinions and view of the world. Once in our mind ideas, the 'facts' become entrenched, and we find it difficult to change them. To some extent, advances in technology are useful, as not only do we have access to Internet viewpoints from across the globe, but satellite TV offers a more diverse spread of opinions than national TV. On many occasions, I have heard the same events reported via a number of satellite channels and been surprised at the differences in viewpoints, even for items that are not political. This is a very positive feature of improved technology.

Not only do the foreign TV channels show that the facts behind the stories may appear very different according to who is presenting them, but also for the very parochial events in our home towns, we often have actual knowledge of the items being reported. When this happens, it frequently becomes apparent that the process of reporting, access to correct sources, time constraints, and the need to sell the papers means we rarely agree totally with the reported version. We should take this as a warning as to the accuracy and impartiality of the global and political statements, as well. Partial or incorrect information may be worse than ignorance.

Deliberately skewed or falsified disinformation is not only routine in political activities, but can be just as significant in items presented as science. Examples run from marketing, where white lab-coated 'scientists' make dubious claims as to the efficacy of a product, to items from high-powered laboratories and individuals. Similar distortions are built into reporting. On one occasion, I was being interviewed for a TV programme, but the TV crew were unhappy that I did not have a white lab coat, as they said it gave more credence to the work. They saw it as a mark of authority. The official 'uniform' is a false image, although we respond to it. In medical situations, blood pressure taken by a nurse in formal uniform invariably produces higher readings, as it raises our stress levels. These are typical cases where technological artefacts distort the information.

Intentional distortions are not rare. I have seen examples where data, known to be incorrect, were not retracted whilst a laboratory was in the process of a grant application, or 'information' released that was consciously intended to divert the efforts of a foreign power. None of this is surprising, since there were politics and careers involved in the examples I have detected. My main concern is to wonder how often I have missed such events and so believed dubious statements.

By contrast, reliance on the media of a single nation may be both unwittingly xenophobic and prejudiced. I recall being abroad in a country hosting an Olympic Games, which I watched on the local TV. I thought the host nation must be excelling, as the only images were of their competitors, but then I realized they were not saying where the athletes had finished in the competition, nor did they mention any winners unless their local nationals were among the medallists. They were catering to, and encouraging, a very parochial mentality.

The same pressures apply to political leaders, who must appear to be strong and maintaining the national interests when dealing with other nations. So they avoid information and decisions that would make them locally unpopular, as this might lose votes and internal influence. Their bias can equally be driven by administrations that filter information input to the front-line politicians. The common cliché is that 'knowledge is power', but political power comes from an electorate that only has carefully selected information (definitely not knowledge).

Failure to exploit resources

Technology can advance via a few key ideas and commercial exploitation, but society in general is far more complex, and progress requires input and benefit to the entire nation. In dictatorships—militarily or religiously governed countries—there are major weaknesses that large sections of the population neither benefit from, nor are able to contribute to, the overall social wealth of the nation. Politically, this is unfortunate, and an inept use of the latent resources of the country. For these extreme examples, their failings are obvious. Nevertheless, even in a country such as the UK, there are a wide range of factors that suggest we are equally guilty in many ways.

The UK is potentially socially rich in having a very diverse multicultural mix that should offer alternative views and contributions to the country, but for various reasons these features either are not exploited, or are counterproductive. Divisions come via class, religion, culture, and race, etc. plus the very obvious difference that the two sexes are not treated equally.

Similar difficulties occur in communication between different religious and ethnic groups and, as always, inability to fully communicate is likely to result in dislike or resentment. Such problems are potentially acute in major cities; an extreme example is London, in which administration and representation involves people using around 300 different languages. Attempting to operate under these conditions with such a diverse population is challenging.

Parliamentary representation and practice

Despite having an elected Parliament, the members are not truly representative of the bulk of the public. Over one third of MPs in 2010 had come from fee-paying schools compared with ~10 per cent of the general population, and some 20 were from just one school, Eton. In some respects, this is not surprising, as Eton has both an excellent academic tradition and a very strong sense of meritocracy driven by both pupils and staff. Such attitudes and selection of ability are unachievable in the majority of schools across the country. Additionally, around 90 per cent of MPs attended university, compared with ~10 per cent of their age group. However, rather than being representative of the diverse range of training across a swathe of universities, the statistics record that (at the time of writing) of the 55 prime ministers since 1721, no fewer than 41 had attended Oxford or Cambridge.

Government is thus dominated by a section of the population that may never have worked outside politics, has no experience in the 'real' world, and has been taught by the same tutors, who similarly may have no practical experience outside academia. The same group inevitably select party candidates and cabinet members from those they are comfortable with. The unfortunate conclusion is that the country survives with a highly self-selective process that does not benefit from the far greater diversity of opinion and background of the nation that is being governed. Equally, the government is unable to closely relate to the needs and attitudes of the majority of the population.

This is information and knowledge rejection on a scale that relates directly to the entire nation. Further, it is a criticism that is not limited to the UK.

Similar types of criticism can be levelled at the governments of most countries, even if the leaders are elected, rather than being hereditary, or imposed by military or dictatorships. For example, the USA has a system whereby the potential presidents are selected by the major parties as a results of state voting, after there has been intense canvassing and self-promotion. The cost of the campaigns is immense, and only those with a very considerable wealth have a realistic chance of having enough media coverage to attract a following. Public records offer approximate numbers for the personal wealth of the candidates, and the additional support money they attract.

Looking at the data, it is not unusual to find that the survivors in the campaigns are multimillionaires or even billionaires. Such levels of wealth may variously have been inherited or the result of major business activities. This may well reveal an aggressive and forceful personality, but it definitely indicates that they are highly unlikely to recognize the living conditions of the majority of the nation. The other problem is that characteristics needed for the campaign trail are not an automatic guarantee that they will be matched by the statesmanship needed to lead the country. A further negative feature of this selection process is that it greatly reduces the possibility of having female or ethnic minority candidates.

It is easy to be critical, but finding better routes to select leaders and governments, and actually implementing them, is a considerable challenge in any relatively democratic system.

The way forward

In this chapter, I have focussed on cognitive rejection and the ensuing problems of failing to learn new ideas, or recognize and appreciate factual evidence. There are downsides to our present ingrained attitudes, although if they result in caution, the reluctance to blindly leap in new directions is reasonable. Nevertheless, the very principle of rejecting new ideas implies that we have limited knowledge and a blinkered perspective. Therefore, if we are to benefit from advancing technology, understanding of a wider information base is needed. In part this must be more scientific than has been fashionable for the training of our leaders who set the directions of the various nations. Unfortunately, science is currently disparaged, even in many major countries—a fact that is incongruous, as the advanced nations are totally dependent on science for everything from communications, power, material goods, food, healthcare, and armaments. Without knowledge in topics as diverse as biology, chemistry, geology, mathematics, physics, zoology, etc. we are totally unable to see a true and total global picture of how our current actions of the dark side of technology are changing the planet in terms of depleting natural resources, producing a diversity of pollutants, and callously causing extinction of other species.

Climate change has been discussed endlessly, and some aspects of altered weather patterns (such as more intense storms) are now being recognized, even by those politically or financially opposed to their possibility, especially if this will alter profits. In reality, our actions to reduce future undesirable climatic effects are minimal, despite many conferences and global meetings. Indeed, some sources of contamination, such as air travel and shipping, are not considered, even though these are immense contributors to the problem. We need major shifts in opinion and understanding which will take a generation or more. Chapter 7 cited the 1962 work of Rachel Carson (*The Silent Spring*). Her message was clear and has been widely accepted as true, but half a century later, the same problems exist (albeit often with newer chemicals and processes).

The planet may support advanced nations for many of you who are reading this, but there is absolutely no certainty that it will be the case within one or two generations.

My hope is that we can gain knowledge, transmit it, and have it appreciated enough to take actions which will offer a long-term survival of humanity. Key to this is a sense of responsibility and an understanding of the science that supports our technologies. This aspect of education needs to be deeply ingrained in our future leaders as the technologies are far more relevant than the classics of past civilizations. I am being critical, not because I am a scientist, but because we seem to have learnt rather little, or very slowly, from historical records. Indeed, many examples of earlier civilizations do not justify the word, as they survived on the basis of slavery and warfare, which were frequently supported by superstitions and different brands of religious intolerance.

My hope is we will focus on education so that our panic reactions against new ideas will be more rational and measured. If we are successful, then we should progress towards a globally interacting society, learn how to stop exploiting and destroying resources for our own instant pleasure and profit, and begin to think of future generations. The difficulties in making this shift in attitudes hits at all levels of society. It will not occur overnight, but it is essential.

15

Hindsight, Foresight, Radical Suggestions, and a Grain of Hope

Civilization and our dependence on technology

In this final chapter I will start with hindsight gained from earlier comments, summarize some key issues, and then attempt to offer foresight on what is needed in the immediate future. My preceding chapters are based on factual data, but here I need to provide suggestions and ideas, even if they are radical ones. My major concern, and indeed the overall message, is that we are not just vulnerable to natural disasters, but because of technology, are increasingly likely to suffer from natural events that previously were irrelevant. My intense concern is that we have, and are still, exploiting and destroying the planet's resources in ways that will lead to our own extinction. It is simplest to consider three types of challenge. For the first category, of, say, major meteor impacts, the events are beyond our control and they are very rare. If they happen, then we may become extinct (that is life). So I will ignore them, as there are more probable contingencies where we can usefully make some preparations.

The second category of natural events (such as the sunspot scenario) are not predictable in terms of date, but they are regular features that now could cause immense damage to humanity because of our total reliance on technology in advanced nations. Here we have a very clear message. We can predict the consequences and, if we have the desire and motivation, then we should already be preparing plans and building the defences that would minimize their impact. The timescale before a significant sunspot emission that hits us is not predictable, but from previous recorded events, it is certain a major one will occur during this century and possibly quite soon.

Our choice here is unequivocal. Immediate preparations would not be excessively costly, and technically they are feasible. Contingency preparations should allow the advanced nations to survive with a reduced death toll and some coherence of continuity. The alternative of inaction and failure to plan and prepare will mean events such as loss of power grids, satellite communication, or both. These are likely to destroy many technically advanced societies. The only slightly positive outcome is that people in underdeveloped societies will probably survive. The continuity of humanity would then be their responsibility, but the changes in world economies would trigger a very retrograde state for global civilization.

My final concern is for our current maintenance of the resources of the planet, plus our ongoing pollution of the land, sea, and atmosphere, and a rapidly expanding population. Although we are totally responsible for all the negative aspects of this situation, we seem unwilling to recognize that they exist. There are many discussions and proposals, but very few real actions, despite the fact that the urgency to counter the downward progress is extremely pressing. We may (or may not) still be able to stop, or even reverse, some of the destruction that we are generating. If we do not act, quite literally within one generation, then by default we are planning the collapse of civilization, and possibly humanity.

A classical scholar might draw an analogy between the dark side of technology and Pandora's Box, which released many evils. This is very apt, as we have embarked on technical progress without understanding future outcomes. If technology continues unchecked, purely for profit and without concern for the consequences, then we are doomed. Nevertheless, Pandora discovered the box had one further item, Hope. This is equally appropriate for our attempts to control the consequences of technology. It is a very small and fragile factor, as it requires us to take actions that are for the global general good, not just for localized affluence and an easy life. My aim is to encourage the growth of this tiny grain of hope, and by active and immediate unselfish actions, we must accept the responsibility of preserving both humanity and the resources and other creatures of the planet. Our little grain of hope has grown in several areas, as in recent years we have at least attacked some of the simpler, less contentious issues. For example, the long-term downsides of atomic weapons are recognized (although it has not stopped nations attempting to build them to demonstrate their skill); asbestos is slowly going out of fashion; CFC pollution has been cut so that the ozone hole may recover to allow it to reform the ozone in the upper atmosphere that provides protection against UV light; we recognize many of the contamination effects of herbicides, etc. and at least some nations have made an effort to limit them. So hope does exist, but it needs encouragement and action for my key topic areas, plus some mechanism to totally change our sense of priorities from consumerism and profits to care for the planet, not just at the present time, but for future generations.

The following paragraphs offer a rapid résumé of dangers where we can have some influence, and where actions that are needed. This is followed by more radical suggestions of attempts to change human behaviour. Failure may mean our extinction, so there is considerable incentive to consider why change is needed.

Hindsight on solar emissions and modern technology

Fourteen chapters ago, I noted how the impressive aurora displays near the poles of the earth have moved from being beautiful entertainment to a potential threat to modern electronic systems. The normal displays are from energetic particles randomly heading out from the sun that are trapped in the atmosphere by the earth's magnetic field. The lights occur over a broad swathe of the atmosphere and represent many hundreds of megawatts of power. The danger for us is that the particles emitted from sunspots are directional, and at far greater intensity than the general background emission. The sun is some 93 million miles away from us; the difference between the random background and a directional beam of particles is some 50,000 times. Further, the sunspot flares can be 10,000 times greater than the normal surface emission. If the energy pulse happens to be directed towards the earth, then the beam from the sunspot may deliver many million times more energy into an aurora electromagnetic storm. This will endanger electronic communications and the electrical power grid networks, together with all associated services we expect in advanced nations. A recognizable analogy is to contrast the light intensity from an old-fashioned light bulb with a highly directional pulsed laser. The former is useful, but the pulse laser will be literally blindingly bright, and do irreversible damage.

Typically, the high-risk countries are at the higher latitudes, and especially in the more polar regions, but truly major aurora activity has been seen as far south as Cuba. Thus even a more limited event extending as far south as the Mediterranean would encompass all of Europe, Canada and the northern USA, Japan, northern China, etc. They would be at the front line of a solar flare sunspot catastrophe. Additionally, sustained power loss to major conurbations will be equally disruptive. Dead satellites imply a very long-term global consequence.

Restoring local facilities would be difficult. A US study of a modest scenario predicted the power grid situation would be out of action for at least a month, but the failure might extend to several years. Personally, I feel the report contained political whitewash, possibly to minimize public concern or panic, as it underplays the scale of the catastrophe. Power and communications loss for even a month could produce a high death toll (measured in millions) and total social disorder, especially if the event were during the winter. Because in the model only the northern states of the USA are assumed to be affected, the erroneous assumption is that the rest of the country could instantly provide food, power, and services.

Foresight for solar flare events is essential as they are certain to happen, and statistically we can expect a truly large one within the relatively near future. Design and investment now may not totally solve the problem, but it should avoid collapse of the regions involved. Additionally, protective measures might minimize terrorist attacks that attempt to trigger similar outcomes.

The second aspect of solar emission events is that electronic systems for communications are far less robust than the high-grade, high-power engineering of the power networks. So destruction of local communication nets is highly likely. Our communications involve satellites and, even if they are temporarily deactivated during a high-power directional coronal emission, their electronic circuitry may be destroyed, or we may not be able to reactivate them. This weakness for us is due entirely to technological progress. So for this eventuality, our forward planning requires alternative communication systems independent of satellites. Their operating performance is excellent, but equally we have put all our eggs in one basket.

Other plans should include ways to override normal optical fibre Internet activity, and block all noncritical usage. Not least as at a time of such a crisis, people will be frantically trying to communicate. Technically, Internet control is feasible, but the design needs to be ready for immediate action. Unfortunately, because this possibility exists, one suspects that many nations and terrorist groups are already experimenting to see if they can take control of Internet traffic in order to disrupt it for political or commercial reasons.

Satellites are so successful that many more are being launched, and this will raise a difficulty a few decades from now. They have a finite life, and once destroyed can break up into many high energy fragments, that in turn destroy other satellites. It is a predictable problem and there may be time and incentive to find a solution. There are already tens of thousands of energetic fragments, so we expect a runaway effect over the course of perhaps a decade. Satellite loss might then be irreversible; satellites will no longer be available as a technology for future generations. Foresight partly exists as operating satellites and the International Space Station are frequently repositioned to avoid collisions.

However, what is needed are ways of removing fragments from the satellite belt. If not, then satellite communication will become ineffective or impossible for later generations. This is a technological challenge that needs imagination. I feel the solution to this problem is far more worthy of fame and a Nobel Prize than many of the more esoteric studies that are currently recognized. This may seem to be a controversial view, but without a solution there will be no way to make progress elsewhere. Therefore, within the science community, as well as industry, the scientific status of the problem should be raised.

The more complex subtext of any discussion in which technology drives disasters that cause the collapse of advanced countries (and very large conurbations) is that there will be less impact on the rest of the world. Therefore, from some perspectives it may even seem desirable. I personally find this extremely worrying, as from various political or religious viewpoints, the collapse of the advanced societies might be welcomed, and it could be triggered by conscious acts using only modest, and existing, technological skills. Current examples of new types of terrorism unfortunately strengthen my argument. Blockage of communication has terrorist potential, and there have already been recent examples of attempts to overload specific Internet sites. The activities were short-lived events, but I suspect they were merely on a scale of training exercises.

Topics where we can control the relevant technologies

In addition to external causes of a collapse of civilization, there are many more insidious ideas that are undermining the future of humanity. Just because they are happening gradually does not mean we can relax, as in many cases we have no way of reversing them, and at best we may only be able to slow down negative factors. Topics I wish to reiterate in this category include our uncontrolled use of natural resources, food production, self-induced failures in healthcare, and the overriding issue of population expansion (which is driving many of our other excesses).

If these are such key issues, then one wonders why we have not already taken more action, or why the excesses are only discussed and campaigned against by a minority of people, who, precisely because they are a small minority, are disparaged as being odd rather than perceptive. Indeed, we are more likely to reject the opinion of a young, casually dressed, bearded environmentalist than that of a well-dressed, mature industrialist. Further, most of us have a parochial view of world affairs, so we only become excited and concentrate on solving local issues, without stepping back to understand the longer-range implications of our actions.

As I have said (probably too often), we are obsessed with material goods and profit; therefore we want new toys, new foods, more travel, better healthcare—and all for less money. The only way we believe we can continue along this route is to have expanding markets (i.e. a larger population). Part of the current commercial strategy is to cause peer-pressure rejection of older items, and continuously throw away functioning items in order to replace them with more fashionable new ones. This waste also applies to the huge amounts of food products that are never used. We want more low-cost production, which is invariably achieved by low wages in underdeveloped countries (the modern equivalent of slavery). We also want more power and minerals (by depleting the mines, forests, and natural resources) and an excess of food (with all the negative aspects of excessive agriculture and overfishing). Consequent on this behaviour is a failure to value our health and education, and therefore we also want more medical care to solve the problems that we ourselves have created.

With knowledge and understanding, we might be able to change this deeply ingrained set of attitudes. I am seeking a totally new epoch in human behaviour where individually we feel, and take, responsibility for actions, and no longer hide behind the view that all decisions must be taken by our elected leaders, or dictators, or that we are merely humans, and life is controlled by some deity (so there is no point in making an effort). My idealistic new approach might alter the redistribution of wealth in society, and we might move away from the current state where, in many countries, 95 per cent of the wealth is owned by 5 per cent of the people. This pattern underplays the diversity in wealth, as a flyer from Oxfam in 2016 says that a mere 62 individuals have the same total wealth as half the world population (i.e. 62 compared with 3.6 billion).

I picked the word 'epoch' deliberately. Our recent technological impact on the planet has already been sufficiently immense that the International Commission on Stratigraphy is currently proposing that we have moved into a new geological epoch, which they wish to term the Anthropocene. (Currently we are in the Holocene epoch.) Their arguments for this new name are that we have made irreversible changes that future geologists will use as markers for this Holocene/Anthropocene boundary, and this will be accompanied by the sixth mass extinction in which at least three quarters of species will become extinct. Unless we are intelligent and make sensible changes, then humans may well be in this vanishing fraction.

The alternative is that we make an evolutionary step in our behaviour and progress along the hominoid chain to a new variant. Most of the earlier branches have only left fragmentary evidence for their existence, although we have more skeletons and information of later hominids (e.g. Neanderthal or Denisovan). Carl Linnaeus updated the name of our species in 1758 and added the Sapiens to call us Homo Sapiens. That was before the Industrial Revolution, so Sapiens may now be slightly more justified in terms of knowledge gained since then. Nevertheless, a new name for a new epoch seems reasonable, but it must be matched with a totally global way of thinking, which is preserving the planet, resources, and species. A potential acronym for these future generations could be made from 'Caring And Scientific Humans', as CASH would appeal to politicians and industrialists, and the name is not trapped in an archaic language.

Foresight, resources, and food

The key word from ecologists is sustainability. This does not mean only growing our own food (although often the taste is worth the effort), but using agricultural, fishing, land, and mineral resources in ways that do not deplete them. It may result in smaller profits and less food, but in a new world model of improved health, via our own efforts, less obesity, and the courage not to fall for glib marketing and wasteful, obsolescent products or throw away food—in that case, smaller profits and production may indeed be adequate. A reminder is that between half and three quarters of various crops are never eaten, so in the advanced nations we have a huge opportunity to economize. If we look at the existing ranges in wages and salaries (even the words define a separation into strata), then in larger companies a factor of 20 or more from top to bottom is not uncommon. Globally, incomes are far more diverse. So particularly for the high-end big spenders, a cut in purchasing power will not undermine the quality of their life, but it would save valuable resources.

Currently linked to food production is an immense input of artificial chemicals, specialized growth hormones, fertilizers, herbicides, antibiotics, and pharmaceuticals, plus recent efforts to breed genetically modified creatures so they are resistant to specific diseases (e.g. as successfully achieved to combat respiratory diseases of pigs). All inputs are costly, and control of the substances can be minimal. Currently, even in the USA, there is often weak control by the Food and Drug Administration. For example, in 2013 they 'asked' drug companies not to sell antibiotics solely to promote animal growth. This seems very naive as the increases in animal size are profitable, so self-regulation by the industries will be flouted. The fact that the excess drugs persist into our food chain is highly undesirable, and fuelling the increases in antibiotic resistance.

In general, such short-term benefits may seem encouraging and receive a good press, but the hidden dark side of many of these approaches has barely begun to surface. We eat the products, and there are enough documented examples of consequent reactions, illnesses, and mutations with permanent genetic effects in humans, that we should be extremely cautious and deeply concerned. These are areas of totally new science.

Historically, watering the Fertile Crescent via canals seemed like a great idea a few thousand years ago, but it eventually led to salinity, low crop yields, and permanent damage. The science behind it was primitive. Modern science is unbelievably more complex, and this means we can never understand more than a tiny fraction of it, and do so from extremely narrow perspectives. The hidden long-term damage may be scaled up in the same way that we have increased the complexity of the initial solutions.

Caution in overexploitation, even for reduced quick profits, is essential if we care about future generations. Unfortunately, drug-driven improvements in animal size, and resistance to specific diseases, may not cause us to just mutate into fatter humans when we eat the food. Film and book plots that have envisaged other changes—from super villain to superhero characteristics, or even mass sterility—may unfortunately be possible unpredicted side effects. In this example, the only safe approach is not to involve drugs that may be mutagenic, but recognizing the possibilities is extremely difficult if, as already has happened, some changes do not emerge until two or more generations later (recent animal studies of Agent Orange show examples in the fourth generation).

The only solution seems to be to minimize all genetic alterations of any items that are in the food chain. In many cases, this is an extremely poor strategy. For example, by genetic engineering of goats, goat milk can contain the antimicrobial protein lysozyme, which occurs in human milk, and which is very effective in preventing diarrhoea. For many poor nations, this is a fabulous advance, as currently some 800,000 children die from it each year (1 in 9 of small-child deaths). Perversity of legislation does not yet allow such modified goat milk to be used, and the tragedy of young children dying continues. Unfortunately, this solution does not involve substantial industrial profits, or there would be more motivation to use it.

The health industry

In the last century, there have been fantastic advances in biology and medicine, developments of drugs, and new surgical techniques. This has produced immense industrial activity in drugs, treatment equipment, and professional expertise in every conceivable area. There is therefore no question that humanity benefits from many advances, but we are discussing the dark side as well, so I will remind you of other factors.

Life expectancy has increased in virtually all nations, not just those with highly advanced healthcare. The USA, where the largest sums are spent, only ranks around thirty-fourth (men) and thirty-sixth (women) in terms of life expectancy of the different nations. The UK, with a free health service, is slightly better at twentieth and twenty-fifth. But both examples show that high spending on healthcare does not necessarily equate to longevity. Indeed, the numbers may actually be distorted by our ability to sustain nominal life in many patients who, in reality, may not to wish to survive in physically or mentally intolerable conditions. This prolongation is currently more likely to happen in richer nations. Life expectancy is strongly dependent on income, lifestyle, and education. Hence, within a single country there are large regional and social differences.

Illnesses such as Alzheimer's and dementia are present in more than a million people in the UK and the USA, and the numbers are obviously increasing annually. At the other end of the age range, the USA has, for the first time, a youth generation that is less fit and athletic than were their parents at the same age. Self-induced diseases related to obesity are still increasing. At least in the UK, a vigorous anti-smoking campaign has made a marked reduction in cancers and other diseases previously seen because of smoking. The education involved in this was far less costly than the health bills to the country (and to the pain and suffering of the patients and their families). Therefore, political actions of this type are effective and should continue with more vigour, as the savings are evident to the politicians. Benefits of better self-care and foresight are obvious in areas such as cancer and obesity, since between one and two thirds of cases are the consequence of activities of smoking, alcohol, drugs, overeating, etc.

There is an emerging pattern of increasing ineffectiveness of antibiotics and other drugs, because the various diseases mutate and survive. This is an essential reason we should minimize drug treatments that are prescribed just because patients request them. Equally, treatments to animals and agricultural products should be more targeted and minimized. Our ability to develop new drugs faster than the mutations of the bacteria and diseases may not be feasible. We must also recognize that healthcare is expensive at every level, from treatment, to facilities, skilled personnel, drugs, and equipment. Although the quality and costs are extremely variable within and between nations, profits are large, and so are the sales and marketing efforts. Publicity is rarely focussed on the improvements we can easily achieve by changing our lifestyles. It is certainly effective, as many self-help groups greatly assist in areas as varied as alcoholism, drug abuse, and relationships with other people. So, in my move to an improved species in the new Anthropocene epoch, a logical advance is to spend far more of the medical budget on education of how to live a healthy life, encourage physical activity, and avoid temptations of overeating, drugs, alcohol, and other excesses. Here I am actually hopeful of some success, as politically this will be a measurable and quantifiable economy for each region. It may also produce fitter, healthier, and happier nations. The only downside might be an increase in population as a result of the pleasures of a fitter life.

Benefits of a smaller world population

The final item on my list of global problems that we need to address immediately is a rapidly increasing population. The present total of ~7 billion is deceptive, as numbers are expanding rapidly in the poorer sections within the richer countries, and in underdeveloped nations. Projections vary, but sensible estimates are ~10 billion by 2050 and 15–30 billion by 2100. Reducing the growth rate, and ideally reducing the present total, must therefore be a global priority. Failure would result in a lack of food and resources for the entire planet. Without reductions, we will have an unsustainable situation, and one can guarantee that plagues and global wars will arise long before 2050. More pessimistic estimates are for even faster growth rates, so conflicts will occur sooner rather than later.

To indicate the scale of changes in recent times, in the period 1990–2010, increases were 62 per cent in Nigeria, 55 per cent in Pakistan, and 42 per cent in Bangladesh. Although more developed, nations such as India or the USA have also seen very large percentage increases. Other countries may initially appear to be more stable, but the numbers can be misleading. For example, the growth rate may look modest if it is, say, 3 per cent per year. For those of us who need simple hints on the mathematics, the guideline is that an apparently small annual increase of 3 per cent produces a doubling of the population within ~25 years. For some countries, the 3 per cent increase can happen as a result of mass migration driven by the exodus of displaced people from war zones, or famine caused by climatic changes and crop failure. There are also deliberate aspects of migration, to change the political or religious balance of small nations. Such strategies have been openly proposed by several religious groups who wish to change the laws and ethos of small nations, basically by high birth rates and immigration, so as to outnumber the host population.

Small population changes can lead to major long-term increases. To understand the significance of a doubling, it is equivalent to adding to the current UK population all the people living in France (or Italy, or any other region with 60 million people). This is not merely a problem of numbers, but on this scale it is also one of very diverse cultural backgrounds and the inability, and indeed no intention, of assimilating or learning the language of the host nation. Nations are fragile, and both factors are standard reasons for political unrest.

This is one of the clearest lessons we can learn from history of earlier civilizations. My example is for the UK, but it applies elsewhere. In the USA, it might be simpler to think in terms of states, not least as several now are multicultural with more than one language in use. For a coherent society, the ability to use two languages is not a problem, but a bonus. Indeed there are active attempts to retain the dying languages of the original Native languages, since there are some 2.4 million people who identify as American Indian or Alaska Native in origin.

Coordinated efforts and population controls are needed globally. Self-regulation and smaller families certainly occur in more educated areas, but as a national policy to drop numbers, perhaps the only serious large-scale consideration has so far been made in China with a policy aiming at no more than one child per family. It has been fairly effective, but not readily applicable to all regions of the world. The 30-year trial shows population reduction is not immediately compatible with an expanding economy. In social terms, a difficulty is that with four grandparents and two parents, the solitary progeny are often spoilt (termed the Little Emperor behaviour). The payback is that when the older generations need help and care as they age, then there is only one supporter for the six of them. Hence I assume the limit will be raised to two children (albeit probably with a sudden population spike). This could offer a stable upper limit.

Such enforced policies will always be unpopular in some segments of the world, but the consequences of them are far less disastrous than the present uncontrolled (or deliberately encouraged) explosion in population growth. Ideally the target should be to *reduce* the world population in a controlled fashion, not as a result of wars, famine, or disease. My enthusiasm for not just slowing the world population, but actually reducing it, is because we should then have sufficient sustainable resources to maintain a high quality of life for everyone. It is clearly a highly idealistic objective, and it will need effort and political charisma from all nations to head in this direction. Politicians, and the public, who recognize the alternatives should already be pressing and implementing the necessary steps.

In Pandora's Box was a grain of Hope. One should therefore cite the fact that on most continents, the overall fertility rate (i.e. the number of children born per female) has declined since the 1950s. Birth control and contraception are opposed by many religious groups (historically so by both Catholics and Muslims). Despite its background, Tunisia should be singled out as a successful grain of hope. In 1957, their first president,

Habib Bourguiba, guaranteed women full citizenship, initial education, and the vote. He also banned polygamy, raised the minimum age for marriage, and allowed women to divorce. Contraceptives came at a later date. This combination of factors has dropped the initial fertility rate from ~7 down to ~2.5. There additionally has been economic growth. So change is feasible if there is political will.

Ideas to produce a revolution in human attitudes

Our challenge is not just to preserve resources, cut the birth rate, and reduce the overall population, but to attempt to modify our inherent aggressive and destructive aspects of human nature. As a species, we have a long history of tribal, nationalistic, religious, and political divisions, which have spawned hate, intolerance, oppression, and wars. Therefore, my suggestion that we change the way humans behave is going to face extreme difficulties! I could just assume it will not happen and accept that within 50 or 100 years we will self-destruct. This would be a pity as, at least on this planet, we appear to be the most intelligent species that has existed so far.

Our world is guided by leaders, and the rest mostly follow without serious dissent. Therefore to make a change we need a far better-educated majority and leadership who all recognize not just local but also global issues. In the current way governments operate, this means we need to move away from the strict divisions of party politics, and the simplistic 'tribal' thinking. Instead, we need a common aim for our own nations, and for the world as a whole. Reality is that we already have a global economy, and therefore all solutions must be considered in terms of global survival, global sustainability, and global welfare.

Words are not enough. Instead, we need tangible changes that will focus on this new direction. Therefore a radical thought is that we reorganize how we operate assemblies such as local councils, Parliament, the Senate and Congress, and the United Nations. So here is a very simple, technically feasible suggestion that would undermine party political divisions and the inherent aspects of separate competing teams, rather than looking at the overall good. Our tribal genetics favour the divisive systems, and just as at sports events, street riots, and in military service, we have a willingness to take sides. Once having done so, we then become irrational in our behaviour if we are with a pack of people with similar objectives. This mob-rule attitude is counterproductive, so if we can undermine it, then there could be progress.

Technology and political seating plans

The UK has a moderately fair voting system to elect members of Parliament—perhaps with a caveat that postal and proxy votes may well be enforced by colleagues or local heads of families or religions. The second drawback of democratic elections is that everyone has a vote, even if they have no understanding of what they are voting for. As ever, this implies we need better-educated electorates. Once elected, only tiny numbers of the representatives actually go to many of the debates (as is obvious from TV coverage). These numbers are totally distinct from those who hear the voting warning bells throughout the building (and elsewhere) and who then rush to vote in person according to the demands of their party. Their decision on the voting is definitely unlinked to any discussion that has been made in the debate, and possibly they may not even have understood the topic. If they disagree with their party, they are unlikely to offer their own opinion, or that of their constituents, if it is bad for their political career. Rarely is a vote defined as a 'free vote', whereas this should always be the case. This is equally true at local council meetings, and I have experience of each political party having a pre-council meeting to decide how they will vote on different topics. The 'debates' within the council meeting are then just window dressing.

For major topics (major as assumed by the political leaders), the House is packed and then we hear some incredibly pitiful school playground confrontational shouting from one side versus the other. Rarely does it seem like a debate, but just ranting with the message that 'We are best, you are wrong!' The most important message that comes across to us, the electorate, is that the content is irrelevant if in some way it can be presented to denigrate the other side. Since the leading members often have similar backgrounds, it is far from clear which of them is truly to the left or right on many issues. Indeed, if they were to behave like independent normal humans, they would never be in such total agreement with their party on every issue.

My minor and modest piece of technological change would be that each member presents their identity card to a random seat number generator as they enter the chamber. They then must occupy that seat. The childish confrontational shouting match would then be very difficult as there could be a total mixture of opinions on adjacent seats. The original two facing sides with a space in between, which was initially designed to stop fighting (for the same reason as the hooks for leaving swords outside of the chamber) should now be irrelevant. This concept is not limited to facing-bench parliaments, but could be equally applied in the semicircle seating in other assemblies such as the US Senate and Congress, or the United Nations (and probably every other national assembly).

At a stroke, this would swamp the opposing-team, mob-rule characteristics, and leave only a common focus on problems that are being addressed. It is difficult to be totally offensive to an opposition if you are sitting mingled in with them. Disagreement is fine, and indeed many problems lack a simple unequivocal solution, so attempts will vary with political instincts, but no longer could the members blame failures as the policies of the other side (there is no other side in my seating system).

The second item of technology (for the UK) would be to dispense with the bell system (including any in local pubs), but when a vote is to be taken it must be made at the seat that has been randomly allocated by the computer, by the person assigned to that seat (i.e. for people attending the debate). Only for the assigned seats would there be a three-button voting system that could be activated. For the other seats, the buttons would be inoperative. The three options would be yes, no, and abstain. It would be totally secret, so votes would be genuine opinion, not forced by the Whips. The use of abstain might well be valuable for those who disagree with the party, but not strongly enough to vote against them on a major issue.

This is not going to solve all political problems, as often there is no solution with laws and political action that can be fully effective. Nevertheless, the random seating would force a very different style of debate that might be far more rational, and stop the confrontational rubbish that we currently witness. It would raise the standard from local politics to a focus on national and global concerns. Adversarial politics inhibits a focus on the outcomes that are for the good of the nation.

The benefits of full equality for women

Progress, understanding, and tolerance are greatly improved in districts, regions, and nations with higher educational standards. Teaching priorities may differ, but a definition of a successful system is one that encourages and enables people to make rational decisions, even if they disagree with their local majority. In a truly educated world, this should not penalize them. Nevertheless, the idea is altruistic and rarely does it occur. It is particularly difficult for women to receive the same level of education as men, even in the higher-status levels of society. Equality in career, job opportunities, and pay are certainly not the same as for men. The rare exceptions effectively underline this disparity. The often unrecognized feature of this is that the losers are not just women, but the entire nation where they live.

Many societies throughout human history have shown, and continue to show, a totally barbaric attitude and treatment towards women. They are denied access to education, receive very severe punishments for crimes or actions for which men in the same society receive no punishment whatsoever, and they are effectively treated as a different species without any of the rights one should accord to them. Current news programmes, TV, and other media give daily examples of these abominable attitudes.

The temptation is to say that it is always driven by religious teaching, but clearly it is somewhat more fundamental in human behaviour, as it is easy to cite examples from many regions, countries, and faiths throughout recorded history. To directly oppose such persecution of women is difficult—especially if it is inspired by religion—as any criticism is seen as interference that will just harden the situation. Therefore a more productive approach is to ask what the society has lost by marginalization of women. The question is relevant even for a society such as the UK, where lip service to equality and voting power exist, but in many careers women are seriously underrepresented, and their salaries are lower.

With a continued focus on the need for knowledge and exploitation of the skills and intelligence in order to build strong and rich societies, it seems blatantly obvious that any society that ignores or degrades half the population is functioning far below its real potential. In fact, failure to educate women and allow them equally opportunities is considerably more serious than halving the pool of potential knowledge and intelligence. Children are critically dependent on the attitudes, information, and skills that they learn in the earliest formative phase of life. Their primary interactions are mostly with mothers (and grandparents), so if the mother is uneducated, then they are all (boys and girls) doomed to mentally develop far below their true potential. There is strong evidence that failure to mentally stimulate children in the first months and years is an irreversible loss for them.

The situation is identical in a husband and wife relationship, as both need to mentally stimulate each other, and this is only feasible if they have comparable education. A high level of education for women, as well as men, is thus a national essential, even if they do not both choose to enter the general work force.

Any political leadership, whether secular or religious, that fails in the education of women is also failing in the development and economic growth of their nation. Advanced nations, whose economies depend on technological skills, are throwing away more than 50 per cent of the wealth and quality of life of their country. The leaders who realize they can double the economic and cultural status of their countries, and then act on it, will be hailed by their nation and go down as heroes for future generations.

In earlier historical times, the superior strength of men was particularly valuable in hunting and manual labour, but this is now irrelevant in most activities. For example, in warfare, the sophisticated equipment is rarely dependent on physical strength, but instead it now needs understanding and control of electronics (e.g. missiles and drones can as easily be targeted and operated by young girls as by mature men). In civilian roles, a very high percentage of men have office jobs requiring computer- or desk-based skills, or work where power is provided by machinery. For such tasks, men and women are equally suited. Indeed, as shown in both the world wars of the twentieth century, even the manual factory and farming skills were competently taken over by women after men had left for the battlefields. A more tenuous possibility is that a higher concentration of women at senior levels of national administration might imply a reduction in the testosterone levels and aggressiveness that has characterized the last few thousand years of human history.

The logic and value of having equality of opportunity, salary, and education for men and women is clear; there is a correlation between gender equality and which are the major successful nations of the current world. True parity does not yet exist, although for leading nations there is a gradual trend that enables women to enter any profession or career, and certainly to rise through the system into key posts in industry and politics. It is still premature to be complacent. Encouragement and pressure on those countries where women are underrated is still an urgent social and political priority.

A final very positive aspect of education and equality for women is that in countries where this is taking place, there are very noticeable drops in the birth rate and improvements in health and longevity.

The educational disaster of war

Whilst failure to benefit from the skills and intelligence of women is frequently a problem arising from religion, it is also made worse by having a male population that is poorly educated. Invariably this means knowledge is replaced by testosterone, with consequent aggression to all and sundry. Historically, there are many examples of knowledge being deliberately suppressed to amplify these traits and increase the military activities. It is not contentious to cite the early European barbarians, or the invaders who rampaged across Asia, or the murderous actions of Crusaders and colonialists.

However, we seem to lack the courage to criticize the same brutalizing attitudes among our front-line troops, or their use of torture, although these acts are still being encouraged in many modern armies of nominally civilized societies. I believe we should oppose such training, not only on moral grounds, but also because the return of military people to normal society is difficult for everyone. There is a difficult challenge in finding a correct balance between military strength and commitment and the destruction of human values.

In the current world, there are a range of nations that have been trapped in civil wars and revolutions (both political and religious) for several decades. Globally as well as locally this is totally disastrous. It means there are entire generations that have had minimal schooling, no social stability, and exposure and desensitization to a culture of killing and torture. The TV images of 10-year-old children, armed with weapons, indoctrinated to kill, is appalling. No matter how soon such wars can be ended, we have generated a legacy to the world where the attitudes of ignorance, intolerance, and war are pervasive, and will continue to be so for many further generations. Warfare is enabled by modern technology, the very dark side of which is damage to the entire fabric of future world civilization.

The two faces of technology

Technology has given us high expectations, an increased human world population, and a demand for a high standard of living. Success has been bought at the price of destroying species and much of the planet in order to produce food, mine ores, and minerals, and to provide water and power to this excess of humanity. My message is thus incredibly clear. Technologies have brought us immense progress and wealth, but simultaneously are sowing the seeds of our destruction. Without a change of intent and actions, collapse of civilization is not impossible. Unfortunately, it may well be in the relatively near future. The grain of hope and a positive view is that some people are now recognizing the dangers, and with immediate actions, we may have the ability to prevent a collapse of civilization and emerge with a better world order.

Despite the evidence that we need to make changes, the requisite actions are likely to be politically and industrially opposed, or slowed down, as they may imply a reduction in growth of markets, particularly for advanced nations. Additionally, those with power and influence are mostly middle-aged politicians and industrialists without a long-term personal involvement. Nevertheless, we need to act now, as the alternative is a collapse of global trade and knowledge, inevitability leading to wars, famine, and disease.

There are perhaps two scenarios that may evolve from our present excesses. The first is that natural events or wars will destroy those of us dependent on technology to survive. This means the end of the advanced nations. This is very bad news for those of us in such favoured positions, but a result that may be welcomed—or indeed, actively triggered—by those in underdeveloped countries.

The second scenario is worse: the dark side of technology may lead to total extinction of our human race by engaging in a global war. In that case, the use of atomic or chemical weapons, followed by global starvation, will combine to eradicate us. Should this happen, then the planet Earth will eventually regenerate, and new creatures will evolve. Perhaps a future sentient being on Earth will detect traces of our existence as a very short-lived species, but it may not realize we self-destructed as a result of our scientific and technological achievements.

Rather than end on a note of deep pessimism, I would like to think that we are an intelligent species that wishes to survive. Therefore, by aggressively pointing out the disaster scenarios we are self-generating, we are taking the first, very tiny step to admitting they exist. Having recognized our mistakes, there is at least hope that we may attempt to correct them. Motivating all of us—from public to politicians and industrialists—with enough urgency to make global changes will be hard, but it must be done. My hope is that you who have read this far will, by definition, have the intelligence, foresight, and drive to not just sit back, but to actively spread this message.

Further Reading

I have mentioned a number of related books that I enjoyed reading, and these are listed below. The scope of *The Dark Side of Technology* is quite considerable, and I have certainly used a vast number of disparate articles, including items in *Scientific American* and *The New Scientist*, general media, and many websites. All have led me into other references. My book is making the point that, not only are there downsides to technology, but also there are some very positive features. Therefore the same topic areas can reveal very different viewpoints. In terms of finding articles and opinions on the various subjects, together with media items, the search engines of the Internet are extremely valuable. I found the more formal documents from governments and major scientific sources are particularly helpful. By contrast, articles that have not had peer review are very variable. Some are informative, some are clearly wrong, and in other areas the personal or political opinions or prejudices override the content. Nevertheless, reading such sites is instructive to see a spread of views and to stimulate ideas. Some items reveal a remarkable lack of understanding of science, so, if in doubt, find corroborative evidence.

Nevertheless, for topic areas of specific interest to my readers, I suggest heading into the items thrown up by the search engines, and not only those that emerge on the first page of the search list. Items buried on the tenth or twentieth page may still be valuable, even if they were not well cited. I am all too well aware that my own scientific papers range from minimal to many hundreds of citations and, as the author, I see little correlation with the quality and importance of the papers and the number of citations. In general, there are more citations for routine items that are discussed by many people, but truly innovative results and ideas have often sat unnoticed for a decade before others started to work on the particular topics—by which time no one uses a 10-year-old reference, and only the work of the newcomers is quoted!

Books that I mentioned are as follows. Note most have appeared in several editions.

Carson, Rachel (1962) *Silent Spring*. Houghton Mifflin, USA; reprinted 2002. Mariner Books.

Fraser, Evan and Rimas, Andrew (2010) *Empires of Food*. Random House, London.

Goldacre, Ben (2009) *Bad Science*. Harper Perennial, London.

Kahneman, Daniel (2011), *Thinking, Fast and Slow*. Farrar, Straus and Giroux, New York.

Klein, Naomi (2015) *This Changes Everything: Capitalism vs. the Climate*. Penguin, London.

Rees, Martin (2003) *Our Final Century? Will the Human Race Survive the Twenty-first Century?* William Heinemann Ltd, London.

Townsend, Peter (2014) *Sounds of Music: The Impact of Technology on Musical Appreciation and Composition*. Amazon Books.

Winston, Robert (2010) *Bad Ideas? An Arresting History of Our Inventions*. Random House, London.

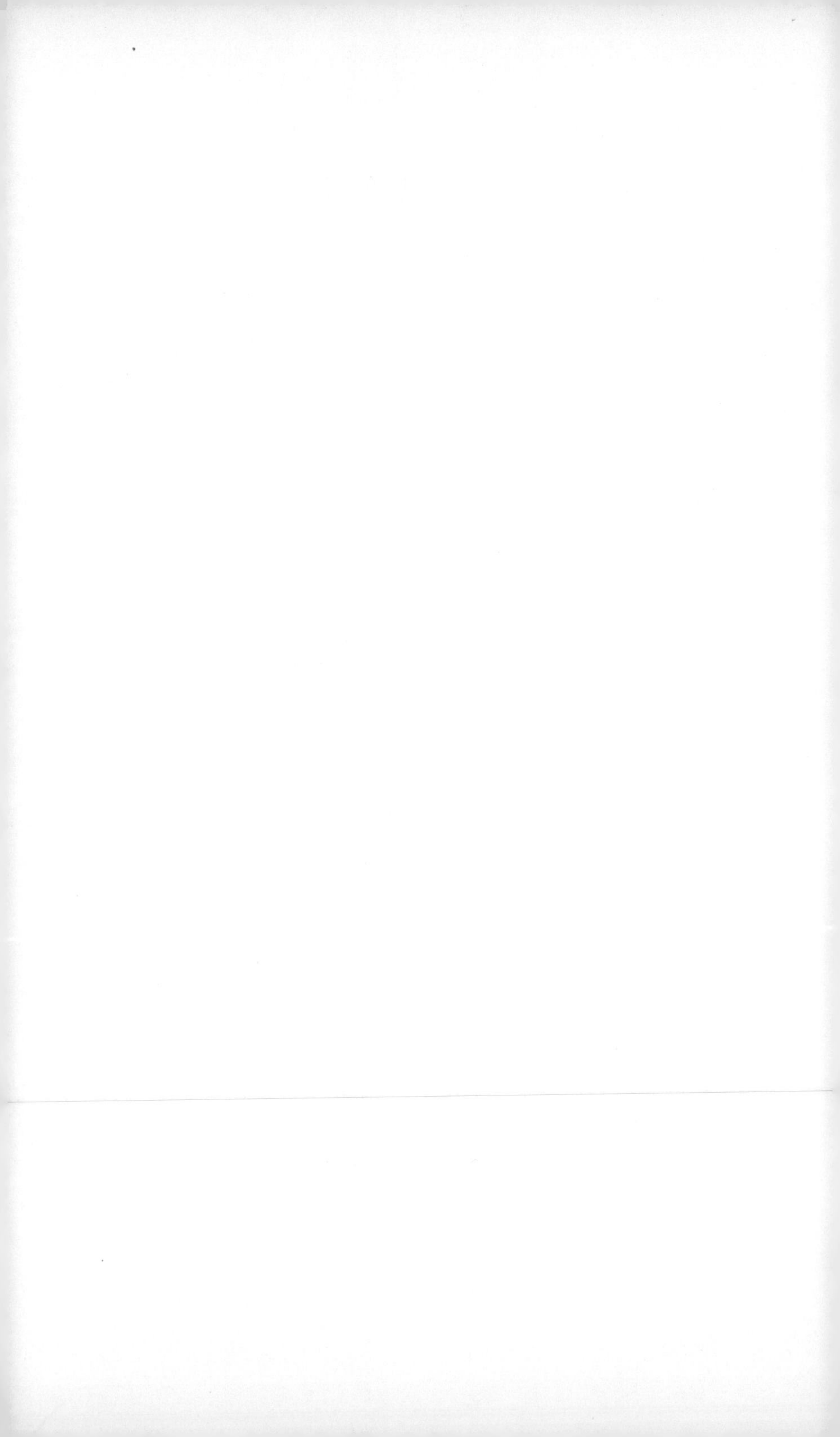